Energy Security and the Indian Ocean Region

First published in 2005, this book is the second volume produced by the Indian Ocean Research Group (IORG). The Indian Ocean Region has become increasingly important to discussions on energy security, not only because of the critical importance of regional states as energy suppliers, but also because of the essential role of the Ocean as an energy route. The main purpose of this volume is to provide an elaborate and critical evaluation of some of these issues and their implications for regions outside the Indian Ocean.

Dennis Rumley is currently Professor of Indian Ocean Studies and Distinguished Research Fellow at Curtin University, Western Australia, and was formerly Australia's focal point to the Indian Ocean Rim Academic Group (IORAG). He was appointed as Vice-Chair of the IORAG at the IORA meeting in Bangalore in November 2011 for a period of two years.

He has been an editorial board member of various international journals and has published 11 books and more than 130 scholarly papers on political geography and international relations, electoral geography, local government, federalism, Australia's regional relations, geopolitics, India-Australia relations and the Indian Ocean Region. He is currently Chairperson of the Indian Ocean Research Group (IORG) and is Foundation Chief Editor of its flagship journal, *Journal of the Indian Ocean Region*. IORG is one of two Observers to IORA.

Dennis Rumley edited a major study on Indian Ocean Security which was launched in Canberra by the Australian Deputy Foreign Minister in March 2013 - *The Indian Ocean Region: Security, Stability and Sustainability in the 21ˢᵗ Century*, Melbourne: Australia India Institute.

He has recently edited *The Political Economy of Indian Ocean Maritime Africa* (Pentagon Press: New Delhi), has co-edited *Indian Ocean Regionalism* (Routledge), with Timothy Doyle, and has had republished as Volume 15 in a new

Routledge Library Editions: Political Geography series his co-edited book with Julian Minghi, *The Geography of Border Landscapes*.

Dr. Sanjay Chaturvedi is Professor of Political Science at the Centre for the Study of Geopolitics, Panjab University, India. He was awarded the Nehru Centenary British Fellowship, followed by Leverhulme Trust Research Grant, to pursue his post-doctoral research at University of Cambridge, England (1992-1995). He has also been a Third Cohort Fellow of India-China Institute at the New School, New York (2010-2012). He was the Co-Chair of Research Committee on Political and Cultural Geography (RC 15) of International Political Science Association (IPSA) from 2006 to 2012.

Founding Vice-Chairman of Indian Ocean Research Group (IORG Inc.), he is the Co-Editor of *Journal of the Indian Ocean Region* (Routledge) and the Regional Editor of *The Polar Journal* (Routledge). He also serves on the Advisory Board of *Geopolitics* (Routledge) and *Journal of Borderlands Studies* (Routledge).

Chaturvedi has served on the Indian delegations to the Antarctic Treaty Consultative Meetings (ATCMs) since 2007 and Track II Trilateral Dialogue on the Indian Ocean (TDIO) among India, Australia and Indonesia. His recent co-authored books include *Climate Terror: A Critical Geopolitics of Climate Change* (Palgrave Macmillan with Timothy Doyle), *Climate Change and the Bay of Bengal: Emerging Geographies of Hope and Fear* (Institute of South East Asian Studies, Singapore, with Vijay Sakhuja).

ARC ACKNOWLEDGEMENT
The editors would like to acknowledge the funding of the Australian Research Council Discovery Project DP120101166: "Building an Indian Ocean Region" for their generous assistance in the conduct and completion of this research, writing and publication 2012-2015.

Energy Security and the Indian Ocean Region

Edited by

Dennis Rumley & Sanjay Chaturvedi

Routledge
Taylor & Francis Group

First published in 2005
by South Asian Publishers

This edition first published in 2015 by Routledge
2 Park Square, Milton Park, Abingdon, Oxon, OX14 4RN
and by Routledge
711 Third Avenue, New York, NY 10017

Routledge is an imprint of the Taylor & Francis Group, an informa business

Publisher's Note
The publisher has gone to great lengths to ensure the quality of this reprint but points out that some imperfections in the original copies may be apparent.

Disclaimer
The publisher has made every effort to trace copyright holders and welcomes correspondence from those they have been unable to contact.

A Library of Congress record exists under LC control number: 2005388664

ISBN 13: 978-1-138-91819-1 (hbk)
ISBN 13: 978-1-315-68867-1 (ebk)
ISBN 13: 978-1-138-91820-7 (pbk)

Energy Security and the Indian Ocean Region

Edited by

DENNIS RUMLEY
University of Western Australia

SANJAY CHATURVEDI
Panjab University

SOUTH ASIAN PUBLISHERS PVT LTD.
50 Sidharth Enclave, PO Jangpura, New Delhi 110014

ISBN 81-7003-302-2

Typeset at
ANJALI COMUTER TYPESETTING
50 Sidharth Enclave, PO Jangpura, New Delhi 110014
Phones: 26345315, 26345713
e.mail: sapub@ndb.vsnl.net.in

Published by
SOUTH ASIAN PUBLISHERS PVT LTD.
50 Sidharth Enclave, PO Jangpura, New Delhi 110014
and printed at
RANG MAHAL
Pataudi House, Daryaganj, New Delhi 110002
Printed in India

Contents

Preface

This book is the second volume to be published by the Indian Ocean Research Group (IORG) and is based on a selection of papers from our Conference held in Tehran in February 2004 held in collaboration with the Iranian Institute for Political and International Studies (IPIS).

Compared with our more wide-ranging inaugural volume, the present book is much more focussed on one of the IORG's particular research themes, energy security, a topic of key policy concern not only in the Indian Ocean Region.

Our third meeting will be held in Malaysia in July 2005 on the theme of 'Sea Lanes of Communication' (SLOCs) in association with the Maritime Institute of Malaysia (MIMA). It is hoped that the fourth meeting will be held in Australia on the theme of environmental security and disaster management.

This year has been important for the IORG for at least two additional reasons. First, we have launched our journal under the expert editorship of Professor P V Rao. Second, we are in the process of developing a lasting partnership with the Indian Ocean Tourism Organisation (IOTO). IORG is especially grateful to the ongoing support for both of these initiatives from His Excellency Said bin Nasser Al Khusaibi of Oman.

For this second volume, we would especially like to record our thanks to IPIS and particularly to its Director-General, Dr. S.M.K. Sajjadpour, and to his many colleagues who were so kind, supportive, efficient and helpful during the Conference, especially Saeed Khatibzadeh and Mohsen Ataelahi. In addition, BHP-Billiton was once again extremely generous and provided major financial support for the meeting. For this we would like to thank

Bruce Becker in London and Siavash Bakhtiari in Tehran. We are all extremely grateful for the very generous hospitality extended to us by H.E. Mr K.C. Singh, the Indian Ambassador to Iran, and for launching our first volume. A special note of thanks should also be directed to Shri Paramjit S. Sahai, former High Commissioner to Malaysia, for his excellent airport diplomacy when, at one stage, all seemed lost.

Apart from our excellent and patient contributors, we would also like to thank several individuals who were helpful in the production and finalisation of this manuscript. In particular, Dennis Rumley would like to acknowledge the very generous Visiting Research Professorship provided by Kyoto University and for the unusually supportive and friendly working environment provided there by Kyoto University Vice-President, Akihiro Kinda, and all of his colleagues. Sanjay Chaturvedi would like to acknowledge the support and encouragement he received from Professor K.N. Pathak, Vice-Chancellor, Panjab University, Chandigarh. He would also like to thank all of the research scholars at the Centre for the Study of Geopolitics at Panjab University, and especially Eva Saroch for continuing to provide valuable research assistance.

Dennis Rumley   Sanjay Chaturvedi
Perth, Australia   Chandigarh, India

Contributors

Aparajita Biswas, Centre for African Studies, University of Mumbai, India

Christian Bouchard, Assistant Professor, Department of Geography, Laurentian University, Sudbury, Canada

Sanjay Chaturvedi, Reader in Political Science, & Coordinator Centre for the Study of Geopolitics, Panjab University, Chandigarh, India

Timothy Doyle, Associate Professor, Politics Discipline, University of Adelaide, Australia

Vivian Louis Forbes, Map Curator, University of Western Australia and Adjunct Associate Professor, Curtin University, Perth, Australia

Girish Luthra, Captain, Indian Navy, and formerly Director Net Assessment in Headquarters, Integrated Defence Staff, New Delhi. Currently, Commanding Officer of a frontline warship of the Indian Navy

A Subramanyam Raju, Director, Centre for Security Dialogue, Hyderabad; Hon. Academic Fellow at Indo-American Centre for International Studies, Hyderabad; and Associate Editor for *Indian Ocean Survey* journal, Hyderabad

Dennis Rumley, Associate Professor, School of Social and Cultural Studies, University of Western Australia, Perth

Vijay Sakhuja, Research Fellow at the Observer Research Foundation, New Delhi, India

Adam J Simpson, PhD Candidate, University of Adelaide, Australia

Swaran Singh, Associate Professor, School of International Studies, Jawaharlal Nehru University, New Delhi and Visiting Fellow with Center de Sciences Humaines, New Delhi, India

CHAPTER 1

Introduction: Energy Security and the Indian Ocean Region

Dennis Rumley and Sanjay Chaturvedi

Introduction

Compared with the first Indian Ocean Research Group -- (IORG) volume, which was a broad overview of some key regional issues (Rumley and Chaturvedi, 2004), this second book concentrates on a central regional policy dilemma, that of energy security. Energy is a key element of national economic development and is therefore an essential component of national security (Keith, 1986). Given the 'underdeveloped' nature of most Indian Ocean states, then this truism has especially significant regional as well as global implications. Discussions over whether 2004-5 indeed represents the third major global 'oil shock' since the second world war, among other things, have focussed renewed attention on questions of energy security. In any event, to a significant degree, the price of energy is associated with supplier stability (Yergin, 1992). Furthermore, from a global perspective, significant investments in the energy sector over the next three decades will be necessary in order to meet growing demand and to avoid a major global economic crisis (International Energy Agency, 2002).

Partly as a result of natural endowment coupled with underdevelopment, while some Indian Ocean states are energy-deficient, several others possess important energy surpluses. In an era of depleting global energy reserves in a context of rapidly increasing demand, the locations of such energy surpluses have become sites for competition between the energy-deficient North and the development-deficient South. This evolving process has a number of other fundamental security implications, not the least of which is the emergence of an international resources diplomacy overlain with the construction of a traditional military security rationale for the 'defence of energy' as part of a new geopolitics of energy agenda. This, in turn, is likely to result in the reconfiguration of some geopolitical alignments that are significantly different from and may be in conflict with those associated with remnant Cold War and other 'traditional' security arrangements, since access to energy supplies and markets will continue to be one of the principal driving forces in national strategies and international relations (Yergin, 1992; CIEP, 2004). Indeed, the question of an "axis of oil" as the basis of a potential "new partnership" between Russia and the United States is indicative of the power of energy security in facilitating geopolitical change (Victor and Victor, 2003).

In the future, the Indian Ocean Region will become increasingly important to any discussion of energy security, not only because of the critical importance of regional states as energy suppliers, but also because of the essential role of the Ocean as an energy route. The principal purpose of the present volume is to provide an elaboration and critical evaluation of some of these issues and their implications not just for the Indian Ocean Region. The volume has five main aims. First, since the term energy security is contested, the volume aims to elaborate and evaluate associated meanings of the concept. Second, some of the concrete Indian Ocean energy security policy issues – for example, transportation – are discussed. Third, a number of case studies of some of the principal stakeholders in Indian Ocean energy security are presented with a view to increasing understanding of their regional goals and strategies. Fourth, regional states possess different strategies to the problem of energy insecurity, and some of these are also considered. Finally, no discussion of Indian Ocean energy security would be complete without a consideration of the implications for environmental as well as human security.

The Energy Security Concept

The case has been well put against the "traditionalists" who want to restrict discussions of security solely to politico-military issues, compared to the "wideners" who want to extend the concept to include economic, social and environmental aspects of security (Buzan, Wæver, and de Wilde, 1998). The present volume belongs unequivocally to the latter orientation because energy is becoming increasingly securitized. However, since the energy security concept itself is contested, some writers discuss it only in the context of problems associated with "security of supply" (CIEP, 2004, 43-4). The writers of two chapters in this volume have taken different approaches to what they understand to be a multidimensional energy security concept. In Chapter 2, Girish Luthra argues that the term energy security is much broader in scope than merely safeguarding the production and distribution of various energy sources. As he points out, in recent decades, energy has also become a key element in international power relations, with Asia having a central role. Luthra places energy security within a conceptual framework that emphasises the protection of interests, however the latter are defined. These interests, in turn, are influenced by considerations related to different types of energy sources. The conceptual framework which is proposed, and which is aimed at the development of a strategy to enhance energy security, is built around two central considerations – the assessment of energy security needs and the adaptation to those assessed needs. Luthra argues that the assessment of energy security needs requires an understanding of the varying expectations of different consumer and producer states, the perception of energy interdependence among consumers and producers and a determination of specific vulnerabilities for suppliers and consumers.

The author suggests that, in essence, an energy security strategy needs to identify a plan for adapting to assessed needs via a combination of mitigation measures – such as sectoral readjustment of energy mix and investment in new technologies – and defensive measures – for example, securing transit routes and creating reserves. Since "the Indian Ocean Region is the key to the emerging geopolitics of energy security", Luthra then applies his conceptual framework to an analysis of energy security within various parts of the region and to extra-regional stakeholders.

Dennis Rumley in Chapter 3 evaluates some of the basic dimensions of energy security both globally and in the Indian Ocean Region as part of a broader systems approach. In addition, he briefly outlines some of the policy responses – spatial/geopolitical, energy transfer, non-spatial and environmental – undertaken by states in their search to maximise energy security. Rumley supports Bracken (2000) in his argument that one of the critical factors in the creation of a 21st century geopolitical paradigm is the redrawing of the world energy map as many Asian states industrialise. Increasing competition for energy supplies and for control over energy routeways increases the probability of inter-state conflict.

On the one hand, while some regional states are 'energy-dependent' – for example, India (coal) and Singapore (oil) – on the other hand, several Indian Ocean states are net energy exporters and, on a global scale, the region contains some very significant 'resource holders'. Apart from the Gulf states for oil and natural gas, these include India, Australia and South Africa for coal, and Australia and South Africa for uranium. However, in terms of the significance of energy in their overall trade profile, five regional states – Australia, Iran, UAE, Indonesia and Oman – could be regarded as "energy-niche economies".

In order to maximise regional energy security, Rumley suggests that regional states should give serious consideration, not only to policies of energy diversification towards renewable sources, but more particularly to cooperative policies which are aimed at the creation of an Indian Ocean Energy Community (IOEC).

The Security Spectrum

It is clear that Indian Ocean regional governments will increasingly develop a range of policies to monitor and enhance national energy security (for example, Australian Government, 2004). Furthermore, in general, from the viewpoint of supply, energy security policy tools will include prevention, deterrence, containment and crisis management (CIEP, 2004. 26). However, the concrete application of energy security thinking in the Indian Ocean Region exemplifies a spectrum of policy responses associated with energy transportation questions, issues associated with trends in energy consumption and problems of maritime boundary delimitation, among others. Since the world's energy supply is likely to remain a prime target for terrorists, maintaining the security of energy routeways is an essential cooperative element in any energy security strategy (Luft and Korin, 2004). Furthermore, in a context of energy scarcity, energy supply locations are likely to become zones of conflict especially where land or sea boundaries are not clearly delimited, irrespective of any idealist views concerning potential cooperation.

In Chapter 4, Vijay Sakhuja is primarily concerned with the complex geostrategic and security requirements of developing and operationalising an Indian Ocean energy security transportation strategy, especially for China, India, Japan and the United States. States will consider all possible means to defend their regional energy security interests, from the deployment of military forces to the application of various aspects of "energy security diplomacy". In addition, the maintenance of the security of sea lines of communication (SLOCs) will continue to be an important element of state strategy. However, with the growth of global terrorism, maritime infrastructure in general, and SLOCs, in particular, have emerged as the "soft underbelly" of state energy security strategy. Sakhuja argues that, notwithstanding their cost effectiveness, tankers, above all, could well be the "achilles heel" of Indian Ocean trade since potentially they could be used as a means of weapons of mass destruction. He also suggests that, arguably, in recent years, the distinction between terrorism and piracy has become increasingly blurred.

All three extra-regional states considered by Sakhuja – China, Japan and the United States – are faced with an energy security dilemma; that is, being directly involved in the legitimate maintenance of the security of vital energy sea lanes, on the one hand, while not being seen to pose any regional threat or being seen to unnecessarily interfere in the affairs of regional states, on the other. India, however, appears to have pursued a more cooperative role, both with the United States and with other regional navies.

Aparajita Biswas argues in Chapter 5 that US and UK aggression against Afghanistan and Iraq was primarily initiated to gain control over energy reserves in the Caspian Basin and the Persian Gulf Region. The aggression was considered necessary by the United States, according to Biswas, because it was in the process of losing its hegemony over oil and gas supplies in the Gulf Region. She argues that this loss of oil hegemony was as a result of a combination of at least four main factors – the nationalisation of production by OPEC members, the Islamic Revolution in Iran, Saddam Hussain's invasion of Kuwait and an increase in regional competition and energy diplomacy on the part of other extra-regional states, most notably China, France and Russia.

Biswas notes the shift in energy demand towards Asia and the importance that the Indian Ocean Region is playing in this shift, both as energy supplier and as routeway. The increased competition for energy supplies, however, provides not only the basis for regional conflict but also provides an opportunity for cooperation, not only in terms of oil and gas pipelines. Biswas points to the emergence of new "energy partnerships" which are concerned with cooperation in terms of investment, technology as well as trade. However, states need to be mindful of the environmental and human rights implications of energy exploitation to meet the shift in demand. Furthermore, Indian Ocean Regional organizations can play a greater role in reducing inter-state tensions and enhancing cooperation to maximise energy security.

In Chapter 6, Vivian Louis Forbes argues that, in order to maximize energy security in the Indian Ocean Region, it is essential to develop strategies for cooperation in managing maritime space and to develop appropriate legal instruments for the delimitation and administration of the coastal zone. In order to develop this argument, Forbes deals in detail with the Persian Gulf as a case study. As he argues, the Persian Gulf, though central to global and regional energy security, is beset with a range of territorial disputes. In some cases, these disputes have inhibited the development and exploitation of energy resources. Furthermore, the security of hydrocarbon reserves is also threatened by regional terrorist activity.

Following an overview of the location of the Gulf in relation to global energy security, the author then discusses in more detail its geopolitical context. A description and evaluation of Gulf practice in relation to the Law of the Sea Conventions is then presented and this is followed by a detailed discussion of maritime boundary delimitation in the Gulf. In the final section of the Chapter, the potential for cooperation and conflict in the Gulf is considered in the context of maximizing energy security. Overall, Forbes argues that, while states may possess a desire to cooperate, large areas of un-demarcated territorial boundaries and the need to determine regional maritime boundaries, especially in the northern and southeastern sectors, are potential sources of conflict. Such issues, in addition to threats from non-state actors, threaten global and regional energy security.

Stakeholders in Indian Ocean Energy Security

Apart from regional energy holders, such as those in the Persian Gulf, the principal stakeholders in Indian Ocean energy security are those developing states in Asia, especially China and India, which possess a rapidly increasing demand for energy, and those energy-deficient states of the North (Europe, Japan and the United States). 'Protecting' the stake is critical to the economic security of these states. In this regard, it has been suggested that one of the several explanatory reasons for the more than 725 US military bases around the world is to afford protection for American dependence on imported oil supplies (Johnson, 2004, 167). However, achieving energy security through the use of military force, as in the case of protecting access to oil in the 1991 Gulf War, should be seen as only one of many potential management strategies. It has been argued that, in the long-term, the domestic application of non-military strategies, such as energy efficiency policies and the use of alternative energy sources, will guarantee energy security (Alagappa, 1998, 686). On the other hand, the geopolitics of energy security for Indian Ocean energy stakeholders must also involve different forms of regional cooperation with neighbouring states as well as a variety of aspects of a resources diplomacy strategy both with near neighbours and especially with distant suppliers. In short, maximising energy security necessarily implies the application of a combination of both a domestic and an international management agenda.

The Chapters in this section deal with three of the Indian Ocean's most important Asian stakeholders – India, Japan and China, respectively. In Chapter 7, Sanjay Chaturvedi critically evaluates the issues surrounding India's search for energy security in an environment characterised by a complex interaction between geopolitics and geoeconomics. He explores the extent to which the geoeconomics of energy security is obliging India to geopolitically reorient itself in terms of its immediate neighbourhood and beyond. Given India's economic growth rate, and since the industrial sector is the largest consumer of energy, the fact that India will be among the top four global energy consumers by the end of the first half of this century requires the development of an appropriate energy security policy. This is especially important since current proven energy reserves will be insufficient to meet the projected demand.

Current policy not only involves accessing new energy supplies and the construction of a South Asian regional energy grid, but also incorporates a resources diplomacy strategy by which Indian public sector companies are encouraged to acquire stakes in foreign oilfields. However, two of the more complex geoeconomic and geopolitical options involve gas pipeline proposals from Iran and from Bangladesh. Clearly, the long-mooted 'peace pipeline' from Iran through Pakistan to India has the potential to contribute to regional peace and stability and the Bangladesh-India proposal also demonstrates the complex interaction between geoeconomics and geopolitics. Regional cooperation on both options not only would help to maximise India's energy security, but would transform present hostile borders into transborder zones of cooperation. To achieve this, however, requires breaking through the "prison of old cartography and territoriality".

Dennis Rumley in Chapter 8 argues that currently Japan is the most energy insecure state on earth and that maximising Japanese energy security remains a significant global issue given the importance of its economy. Due to its paucity of energy resources and the regional competition for energy supplies in Northeast Asia, Japan is pushed towards a greater dependence on untapped energy reserves in the developing world, towards a greater reliance on nuclear energy and towards a much greater investment in alternative energy sources and energy technology.

Energy dependence has meant that Japanese policy-makers have been especially concerned with issues associated with "supply risk" and thus nuclear power is regarded as an attractive long-term option because it can be seen as a "semi-domestic energy source". It is also argued that the use of nuclear energy is one means of mitigating climate change.

Japanese resources diplomacy with important Indian Ocean Region energy suppliers is another important policy element. In addition, like other extra-regional stakeholders, maintaining the security of Indian Ocean energy routeways is critical. Notwithstanding regional energy competition, Japan has been exploring energy linkages with Russia, especially in terms of possible oil and gas pipelines. Furthermore, greater regional cooperation, especially with South Korea, is a prospect for the future. However, perhaps the most geopolitically sensitive alternative energy security issue is Japan's involvement in the International Thermonuclear Experimental Reactor (ITER) proposal, which, if located in Japan, would generate large quantities of cheap power but would also turn it into a virtual thermonuclear superpower.

In Chapter 9, Swaran Singh argues that, in the next two decades, China's already considerable demand for energy will dramatically increase along with its dependence on energy imports resulting in increased regional and global energy resources competition. However, as the author points out, China's energy deficit cannot be considered simply in the same terms as that of Japan, for example, since it is not import-dependent in the same sense. Rather, China's energy deficit is extremely complex and much more nuanced and structural and arises primarily as a result of a flawed national energy policy. Consequently, China represents a case of the simultaneous co-existence of an energy surplus and energy deficit.

One of the outcomes associated with this situation is that China has been taking a renewed interest in Indian Ocean energy holders and in the maintenance of secure energy supply lines. These developments, in turn, have become a cause for concern among some regional states. From a Chinese perspective, though, increasing its engagement with energy-rich Indian Ocean states should be viewed as no more than an extension of its new resources diplomacy strategy that began to emerge from the early 1990s. However, Chinese Indian Ocean engagement must be considered on the basis of a needs-based *as well as* an ambition-based rationale. On the one hand, in the short-term, China's increasing global economic interdependence implies the adoption of a cooperative security strategy. On the other hand, appropriate management of resources competition is essential to avoid potential energy conflicts in the future.

Approaches to Energy Insecurity

Notwithstanding the appropriate management of competition for energy resources and the emergence of cooperative approaches, the combination of increasing global energy demand, growing energy shortages and proliferating energy ownership contests may well be contributing to an "emerging landscape of conflict", one of the ultimate potential outcomes of which is an 'energy war'. For the most part, however, states of the Indian Ocean Region and elsewhere are more likely to pursue negotiated rather than conflictual outcomes to the challenge of energy insecurity (Klare, 2002, 23). There exist at least three clusters of policy approaches to this issue that can be identified – a demand strategy, a supply strategy and a cooperative strategy. The essential goal of the demand strategy is to implement a range of conservation policies and practices designed to reduce demand and thus enhance environmental security. One of the additional outcomes of this strategy is that it will reduce the relative cost of energy per unit of output, and, to a degree, this has already been happening with oil – that is, since the first oil shock in 1973, the world now uses about 40 per cent less oil for each unit of output (McRae, 2004).

A supply strategy, though equally multifaceted, is funda-mentally concerned with the 'domestication' of energy supplies. This involves locating new domestic supplies of energy, using alternative energy sources, reducing imports of energy and/or finding alternative more secure routes for energy transportation. Using alternative domestic energy sources in this strategy will likely have a contradictory impact on environmental security. On the one hand, it will accelerate interest in the use of renewable clean energy sources. On the other hand, however, such a strategy will almost inevitably lead to a regional increase in the use of nuclear energy and thus an increase in the demand for uranium from relatively stable and secure Indian Ocean states such as Australia and South Africa.

A third policy approach to the problem of energy insecurity represents the opposite of energy conflict and involves a range of strategies including bilateral energy cooperation, regional cooperation and the development of a regional energy security community with its own well-developed and secure linkages. Among other things, such secure linkages may well emerge out of an increase in regional "pipeline diplomacy", since pipeline sabotage is potentially a significant security problem in the Indian Ocean Region (Luft, 2005).

A. Subramanyam Raju, in Chapter 10, deals with this third strategy, and, more especially, with the process of energy cooperation in South Asia. On the one hand, to date, many South Asian inhabitants have relatively low levels of energy consumption and many lack basic energy needs. On the other hand, current regional economic development strategies, population growth and economic reform are collectively contributing to a rapid increase in the demand for more energy in South Asia. However, the inadequate exploitation of significant energy reserves has ensured that most South Asian states must remain dependent on energy imports. One important policy approach to the resultant problem of energy insecurity has been seen to be regional energy cooperation. This was formally raised at a workshop in Dhaka in 1998, and, among other things, the setting up of a Regional Power Grid was proposed. In addition, as part of a vision for an integrated regional policy, it was suggested that a regional energy database could be created along with common energy codes and technical specifications.

Following a discussion of India's energy security scenario, Raju evaluates its cooperative policy strategy in terms of overseas investment and regional bilateral cooperation. He also evaluates the options for the Iran-Pakistan-India natural gas pipelines proposal. Raju calls for the establishment of a South Asian Forum on Energy (SAFE) that would be designed to promote regional energy cooperation and undertake research to improve energy efficiency. He concludes that political relations would improve as a result of greater energy cooperation.

In Chapter 11, Christian Bouchard discusses quite a different approach to energy insecurity when he analyses the energy challenge for the small island developing states and territories (SIDS) of the southwestern Indian Ocean Region. Since the SIDS are relatively isolated, are environmentally fragile, possess small vulnerable economies with limited natural resources, yet have untapped potential for renewable energy, they are currently obliged to be dependent on imported fossil fuels. Bouchard provides the first regional assessment of energy security in the SIDS and proposes a scenario that is designed to gradually lead to regional energy self-sufficiency. Overall, he argues that, since the SIDS have developed their own energy infrastructure and thus are effectively "closed systems", that local approaches to energy insecurity are preferred to more broadly-based regional cooperation strategies.

Bouchard determines the energy balance and degree of energy dependency for each of the SIDS and shows that local energy never contributes more than 25% of primary requirements, except in the Comoros (75%). On the other hand, given the high cost of imported fossil fuels creates a competitive opportunity for local renewable energy technology. Technological developments, especially in the area of wind turbines, reinforce this prospect. In addition, geothermal, solar and other renewable energy sources are available and are being considered as serious policy options. Moving towards energy self-sufficiency, however, will require considerable infrastructural changes as well as a strong political, financial and technical commitment. As a result, SIDS energy self-sufficiency, if it does eventuate, will not be reached for several decades.

Environmental Security and Energy Security

To a significant degree, 'traditional' Cold War concepts of energy security still dominate the foreign policies of many states. However, other factors have become increasingly important in the 21st century, especially the impacts of globalisation, the privileged status given to the market and the continued emergence of 'non-traditional' security considerations associated with regionalism and with environmental and human security. It is evident, for example, that, in the Indian Ocean Region, there is a controversial relationship between energy resources and standards of living and human development for those states that are supply holders. In addition, central to any consideration of the relationship between environmental security and energy security is the question of 'energy sustainability', the key elements of which that have to be reconciled include sufficient growth in energy supply to meet need, energy efficiency and conservation, public health concerns and the protection of the biosphere and the prevention of more localised types of pollution (Brundtland, 1990, 213).

The debate over climate change is central to energy sustainability concerns and the relationship between carbon dioxide emissions and climate change is much more than a function of cost. However, in the United States, for example, energy is very cheap because its price does not reflect any environmental costs and their implications for human health (Hawken, 1993, 76). To a significant degree, this issue lies close to the heart of what might be called an 'energy-environmental security dilemma'. On the one hand, taking a purely market view from the perspective of hydrocarbons, only a price shock is likely to encourage an increase in environmental security. On the other hand, such a price shock may also contribute to the adoption of alternatives which could turn out to be more environmentally insecure. The first option could mean the adoption of alternative clean renewables. The second option could mean an increase in nuclear energy with all of its attendant potentially problematical environmental consequences.

As is well known, combating human-induced climate change requires states to reduce the carbon intensity of their economies, and it has been argued that, by 2025, G8 members should generate at least 25 per cent of all of their electricity from renewable sources (ICCT, 2005). However, to some commentators, climate change is happening so rapidly that a 2025 deadline will be too late to avoid its most serious implications, and that, in fact, the only major solution to solving global warming is for the rapid expansion of nuclear power (McCarthy, 2004). This argument is especially relevant for the ten states possessing the world's largest absolute volumes of carbon dioxide emissions – USA, China, Russia, Japan, India, Germany, Canada, South Korea, Italy and France. In 2004, of these states, only China, India and South Korea were projected to have significant increases in nuclear power over the next two decades, while, by 2025, both Germany and Italy were projected to have no nuclear power (US Energy Information Administration, 2004, 170). However, the 2004-5 'oil shock' has prompted the United States to seemingly reassess its position on nuclear energy, and, in a speech in April 2005, the US President declared that, "a secure energy future for America must include more nuclear power" (CNN, 2005).

Apart from the current small amounts of nuclear power generated by India and South Africa, one possible regional scenario from this perspective of climate change is that other Indian Ocean states will also seriously consider the development of nuclear power programmes. Iran has already indicated its intention in this regard. Furthermore, Indonesia, even though in favour of increased regional cooperation in the field of renewable energy, continues to debate the development of nuclear power, an option which it shelved in 1997. As it stands, the current plan is for Indonesia to begin the construction of its first nuclear power station in 2010, with a view to commencing nuclear power production in 2016 (Agence France-Presse, 2005).

Timothy Doyle in Chapter 12 critically reviews the energy-environment interface in the Indian Ocean Region via the adoption of an inclusive definition of environmental security. He points to the need for Southern environmentalists to shape their own green agendas and to incorporate those that are supported by a majority of Indian Ocean inhabitants. Doyle argues that, just as many parts of the industrialised North have come to realise that the nuclear energy option is unsustainable, the Indian Ocean Region is undergoing rapid nuclearisation, with many states embracing the technology for the first time. As a result, the Indian Ocean is becoming a "nuclear ocean". Doyle recasts the nuclear question in the Indian Ocean Region as an environmental security issue, using, in part, the Barnett and Dovers framework. He focuses on two key issues of the nuclear fuel cycle – the unintended release of radioactive materials from reactors and the problem of the transport and storage of nuclear waste.

After briefly evaluating the history of nuclear issues and movements in the "minority world", Timothy Doyle reviews the five principal anti-nuclear campaigns currently in progress in Australia and suggests that the next level of campaign is likely to be aimed at the proposed transport of nuclear waste. In terms of land transport, for example, Doyle refers to the "battles for Gorleben" in Germany. For the Indian Ocean, however, it is clear that an increasing volume of waste is passing through the region, primarily aimed for Sellarfield in England and Le Hague in France.

In contrast, Adam Simpson in Chapter 13 argues that the maximisation of energy security via a liaison between authoritarian regimes and multinational corporations can have a detrimental impact both on human rights and on environmental integrity. He uses earth rights – the nexus between human rights and environmental protection – as an organising concept for the evaluation of energy projects undertaken by the Burmese and Thai governments in conjunction with Northern multinational companies. Simpson's principal aim is to evaluate the detrimental impact of the Yadana gas pipelines and other large-scale energy projects through the dual prisms of energy security and earth rights.

As Simpson demonstrates, while promising minimal environmental impact at the outset, the Yadana project actually destroyed large tracts of pristine ecologically-vulnerable forest in Thailand during its construction and passed through several protected areas. In Burma, on the other hand, the military repeatedly used forced labour to work on the pipeline, and, in some cases, local villagers were required to function as human mine sweepers. The Burmese military also undertook systematic human rights abuses against local ethnic minorities, including many instances of rape and sexual violence. Legal redress in the US courts, most notably in "the Unocal case", has important implications for the role of US energy multinationals in dealing with repressive regimes. However, as Simpson points out, only continued pressure from foreign governments, international human rights groups and NGOs will help ensure human security to those who happen to live in the vicinity of large-scale energy projects in authoritarian states.

References

Agence France-Presse (2005), "Indonesia agonises on nuclear switch", *The West Australian*, 20 April, p. 28.

Alagappa, M., ed. (1998), *Asian Security Practice: Material and Ideational Influences*, Stanford University Press.

Australian Government (2004), *Securing Australia's Energy Future*, Canberra: Department of the Prime Minister and Cabinet.

Bracken, P. (2000), *Fire in the East: The Rise of Asian Military Power and the Second Nuclear Age*, New York: Perennial.

Brundtland, G. H. (1990), *Our Common Future*, Australian Edition, Melbourne: OUP.

Buzan, B., Wæver, O. and de Wilde, J. (1998), *Security: A New Framework for Analysis*, Boulder: Lynne Rienner.

Clingendael International Energy Programme (CIEP) (2004), *Study of Energy Supply Security and Geopolitics*, Final Report, The Hague.

CNN (2005), "Bush urges more refineries, nuclear plants", CNN.com, 27 April.

Dupont, A. (2001), *East Asia Imperilled: Transnational Challenges to Security*, Cambridge University Press.

Hawken, P. (1993), *The Ecology of Commerce: A Declaration of Sustainability*, New York: HarperCollins.

International Climate Change Taskforce (ICCT) (2005), *Meeting the Climate Challenge*, London: The Institute for Public Policy Research.

International Energy Agency (2002), *World Energy Outlook*, Paris: OECD/IEA.

Johnson, C. (2004), *The Sorrows of Empire: Militarism, Secrecy, and the End of the Republic*, New York: Henry Holt.

Keith, R. C., ed. (1986), *Energy, Security and Economic Development in East Asia*, New York: St Martin's Press.

Klare, M. (2002), *Resource Wars: The New Landscape of Global Conflict*, New York: Holt.

Luft, G. (2005), "Pipeline sabotage is terrorist's weapon of choice", *Energy Security*, 28 March, Institute for the Analysis of Global Security.

Luft, G. and Korin, A. (2004), "Terrorism goes to sea", *Foreign Affairs*, Vol. 83 (6), pp. 61-71.

McCarthy, M. (2004), "Only nuclear power can now halt global warming", *The Independent*, 24 May.

McRae, H. (2004), "It shouldn't happen now, but there's nothing like the black stuff for leaving us in the mire", *The Independent on Sunday*, 20 June, p. 14.

Rumley, D. and Chaturvedi, S., eds. (2004), *Geopolitical Orientations, Regionalism and Security in the Indian Ocean*, New Delhi: South Asian Publishers.

United States Energy Information Administration (2004), *International Energy Outlook 2004*.

Victor, D. G. and Victor, N. M. (2003), "Axis of oil?", *Foreign Affairs*, Vol. 82 (2), pp. 47-61.

Yergin, D. (1992), *The Prize: The Epic Quest for Oil, Money & Power*, New York: Simon and Schuster.

THE ENERGY
SECURITY CONCEPT

CHAPTER 2

Conceptualising Energy Security for the Indian Ocean Region

Girish Luthra

The term "energy security" is a much broader issue than the implied meaning of defensive and protective measures to safeguard production, embarkation, transport, disembarkation and distribution of various sources of energy. In recent years there has been an increasing focus on this subject, particularly since both developed as well as emerging economies have become increasingly reliant on energy for development. Energy has also been a key attribute of changing international power equations during the last fifty years, and continues to be a dominant theme in the emerging geopolitical environment.

Asia is central to the current global debate on energy security due to the rapidly changing supply and demand sides of the equation, which inevitably impact upon the nature of competition. Rapidly rising demand as well as the emergence of new sources of supply in the region, in particular Central Asia, have forced many countries, both inside and outside the region, to reexamine their respective strategies. It is true that the new competition in the region is not insular and is inextricably linked to the global equation, yet the salience of intra-regional dynamics related to energy has very substantially increased. The International Energy Agency (IEA) estimates that world energy consumption is likely to grow by 2.3 percent to 2020, whereas consumption growth in the

Asian region is likely to be between 4 to 6 per cent. It is also estimated that Asia is set to overtake North America as the leading energy consumer in the next 10 to 15 years. During the same period, growth in demand is likely to be negligible in regions like Europe and Africa, and marginal in North, Central and South America. Despite increasing emphasis on diversification to other forms of energy, the share of oil and gas in total global energy demand is expected to rise from about 60 percent to 68 percent by 2020.

While concerns about energy security have led to subtle posture and policy changes in respect of many countries, the issue is viewed very differently by different states, based on their perceived strategic interests and relationships within the overall grand strategy. Further, different organizations look at energy security differently, coloured as they are by their respective organizational perspectives. Diverse approaches link the subject with different considerations, and many countries have found it difficult to come up with a coherent and integrated energy security strategy.

What is "Energy Security"?

Before addressing this question about energy security, it may be prudent to revisit the term "security" itself. Derived from the Latin word securitas, it refers to tranquility and freedom from care; yet, this meaning does not seem to convey all that falls under the ambit of "security." Distinctions have been made between so-called traditional and non-traditional understandings of security. Absence of risk and fear has been offered as other theories, but such theories have not been able to capture the entire meaning of the word. Further, there are categorizations such as "state security," "international security," "comprehensive security," "common security," and so on. The underlying theme in all of these concepts of security is that it has multiple levels of interaction and that it is intended to safeguard interests, whichever way they might be defined. It is with this under girding interpretation of security that "energy security" can be examined, and placed in a conceptual framework.

Energy Security itself implies safeguarding interests that are influenced, directly or indirectly, by considerations based on different types of energy sources. It is generally understood to be linked to the development process, ensuring sustainability over long-tern. Essentially, it comprises elements of national security, economic security and environmental security. Unlike some other forms of security, it is more "vulnerability based" rather than "threat based." It also implies making strategic choices in political, economic, social, military and environmental realms, and comprises elements of both collective as well as national security. Its assessment leads to explicit or implicit strategies to prepare for resource/supply disruptions, maintain access/supply at reasonable prices, and adapt technological measures and consumption patterns to promote sustainable development with least damage to the environment.

Devising a Conceptual Framework

The development of any strategy to provide better energy security for the future depends largely on the contextual framework in which energy-related interests are perceived. An analytical model that allows a comprehensive examination of energy security can be useful for the purpose. Figures 2.1 and 2.2 illustrate a model that seeks to provide a conceptual framework for examining energy security in depth.

A comprehensive assessment of energy security needs is essential before examining issues related to policy and strategy (Figure 2.1). This assessment process can be attempted under three broad headings – energy expectations, energy interdependence and energy vulnerabilities.

Energy Expectations

Expectations from energy vary considerably among different countries. Based on the respective energy mix, the focus also differs although oil and gas are currently high on the agenda of many states from the viewpoint of expectations. The sector-wise distribution of economic activity forecast, too, has an impact on overall expectations. While suppliers have major expectations from markets (including routes, price stability and so on), demand countries have major expectations from existing and emerging areas for sustained supplies. Consumer markets also have expectations from different sources for "ramping up" production and supply to ride turbulent periods (commonly referred to as "swing production").

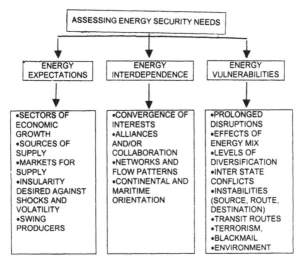

Figure 2.1. Energy security: needs assessment.

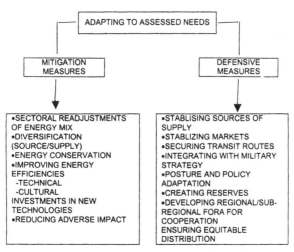

Figure 2.2. Adapting to assessed needs.

Energy Interdependence

Perceptions of energy interdependence and convergence of interests seek to link national security with collective security. Producers need a steady stream of revenue for development while consumers focus on stable supplies. Alliances and/or collaborations at the political level can be gainfully employed to enhance mutual interests and strengthen respective "firewalls" through shared concerns, though implementation of this process has numerous practical impediments. Assessments related to desirable as well as likely 'flow patterns' and 'networks' of supply and distribution are also critical in assessing energy security needs. Shared interests in a flow pattern which may have a continental orientation (pipelines over land) or a maritime orientation (undersea pipelines and shipping mode) can lead to different forms of interdependence and the emergence of a new architecture at the geopolitical level.

Energy Vulnerabilities

Assessment of energy vulnerabilities, specific to each supplier or consumer, is fundamental to a comprehensive analysis of energy security. Vulnerabilities may vary depending upon the nature of energy mix, risk assessment of disruption, levels of diversification and assessed probabilities of both inter-state and intra-state conflict. Further, vulnerabilities are linked to existing and likely instabilities in areas of energy source, transit routes and destination. The potential for fundamentalist threats and terrorism add to vulnerabilities, and limitations of security and economic guarantees with regard to new forms of threats enhance perceptions of vulnerability. Coercion and blackmail, which are consistent themes in the current world order, also need to be taken into account when examining energy-related vulnerabilities. In terms of environmental degradation, collective vulnerabilities are more pronounced although lack of concerted action on environmental issues can lead to increasing polarization and unequal emphasis on such vulnerabilities.

An Energy Security Strategy essentially outlines the plan for adapting to the assessed needs. Such an adaptation can be examined under two broad headings – mitigation measures and defensive measures.

Mitigation Measures

These measures include readjustments in order to reduce vulnerabilities to the extent possible. These include sectoral readjustments of energy mix, diversification in respect of sources of supply/markets, improving efficiencies in energy consumption, investments in renewable as well as promising alternate sources of energy, and initiation of processes to reduce the anticipated ill-effects of energy usage (Figure 2.2). Mitigation measures, too, have significant hurdles and constraints, which, in turn, impose limitations on strategy development.

Defensive Measures

These include proactive measures to safeguard assessed energy-related interests. These are a combination of diplomatic, economic, military and informational initiatives which include stabilizing supply sources and markets, securing transit routes, integrating requirements of energy security into the overall military strategy, and policy adaptations to enhance leverage and protection.

Energy Security and the Indian Ocean Region (IOR)

The conceptual framework outlined above can be used to examine energy security specific to each country. Since the IOR is a very large region, it is appropriate to address the issue by examining aspects specific to each of the different sub-regions of the IOR, as well as other sub-regions with an important energy interest in the IOR.

Middle East and the Persian Gulf

Oil and gas energy use is the dominant theme and given that oil and gas energy use is expected to rise by about 75 percent in BTU between 1997 to 2020, the Middle East region is the most important region from the point of view of meeting future supplies. It has about 65 percent of the world's proven oil reserves and 34 percent of its gas reserves. Over 90 percent of these reserves are in the Persian Gulf region. The Gulf in expected to produce 56 percent of world oil exports in 2020, up from 43 percent in 1995. Despite the emergence of new sources of supply, the importance of the Persian Gulf is set to grow in the next 10 to 15 years.

Countries in the Middle East have expectations that oil and gas reserves, particularly low cost-high quality attributes, would contribute to overall economic development, an aspiration that has been somewhat elusive in the past. Major challenges related to structural adjustments and demography are likely to be brought into balance through steady revenues. Difficulties experienced in the past due to cycles of "oil boom" and "oil crash" and problems of under-investment in energy infrastructure have led to expectations that the emerging markets of Asia will provide the requisite stability. Gas revenues are expected to substantially enhance oil revenues, and some countries also expect to be the link route for connecting Central Asian oil and gas to markets in the Indian Ocean Region. For the foreseeable future, "swing producers" will continue to be from the Persian Gulf region, raising expectations from the leverages generated by increasing import dependency of many countries on the region. Technical and financial assistance is considered crucial by many to meet long-term aspirations based on oil and gas revenues.

Cooperation in energy exports is also seen in terms of keeping intra-regional and internal conflicts at bay. Political, military and economic alliances continue to dominate with energy as the key driver. While a maritime orientation has been of primary focus in the past, some countries of the region, particularly in northern Persian Gulf, are now placing more emphasis on a continental orientation. Some estimates suggest that, by 2020, LNG imports in the Asia-Pacific are expected to be over three times more than pipeline imports, which, in turn, implies an increased importance of the IOR as a transit route.

The suppliers in the region look at respective vulnerabilities very differently. These relate to intra-regional disputes, geographical location and size, limited progress in terms of diversification in other sectors of economy, choke points (Straits of Hormuz), internal security, terrorism and foreign military presence. Vulnerabilities are also seen in terms of price volatility, production quotas and sanctions. The new portrayal of the region in the context of fundamentalism and terrorism has only increased energy-related vulnerabilities.

The countries of the region are clearly moving towards a new energy strategy, albeit at a different pace. Joint ventures, foreign investments, curtailing budget deficits, and maximizing revenues, while conserving for the future, are some of the measures being adopted by a few countries. Reliance on security guarantees by external powers and cooperation in keeping energy trade routes secure are seen by many as essential for strengthening energy security.

Central Asia

Though no Central Asian country is an IOR littoral, the region has significant relevance in terms of energy-related geopolitics of the IOR. While the region has limited proven oil and gas reserves, its potential for becoming a key supplier is many times greater.

Energy sources of the region are generally concentrated around the Caspian Sea region. While oil reserves are mainly in Kazakhstan and Azerbaijan, gas reserves are mainly in Turkmenistan, Kazakhstan and Uzbekistan. The region, which was a net importer until the early 1990s, has seen a gradual rise in the production of oil and gas. Exports are expected to rise considerably in the next 10 to 15 years, with about 3 million b/d of oil and about 1 bcf/d of gas being exported in 2015. These figures are very small when compared to the Middle East, yet the region has gained in importance on two counts – its gradually increasing exportable surplus and its promise of providing an alternate source to markets that are yearning for diversification. A balance is also being sought in supplying to mature markets to the west and the emerging markets of Asia.

The countries of the region have major expectations from their reserves and exportable surplus but face numerous technical and political difficulties. Leverages generated by its portrayal as a viable alternative energy supplier have raised overall expectations. While an operational pipeline network to Russia continues to be used, other attractive markets are also being explored. The task of broadening markets is problematic, but oil pipelines connecting to Black Sea terminals and gas pipelines in all four directions - north, east, south and west – either under construction or being examined for feasibility. These include:

➤ Pipelines form Turkmenistan to Russia via Kazakhstan and Uzbekistan.
➤ A Baku-Ceyhan pipeline from Azerbaijan via Tblisi in Georgia.
➤ A Kazakhstan-Xinjiang pipeline and a Turkmenistan-North China pipeline.
➤ Pipelines to Iran and another pipeline to Afghanistan, Pakistan and India.

Perceptions related to converging interests with other countries vary significantly. Political realignments and economic collabora-tions have led to the euphemism "pipeline politics." Access to the Indian Ocean via the Gulf region offers promise, but the dominant view is supportive of a continental orientation with collaborations in setting up a diversified pipeline network.

The countries of the region have energy vulnerabilities dictated by geography, the influence of major powers and neighbours, the security of the pipeline network including transit through unstable regions, and a lack of diversification and limited freedom of manoeuvre with respect to the routing of pipelines. Land pipelines are also subject to the vagaries of complex and changing political relationships, which, in turn, add to vulnerabilities. Fundamentalism and terrorism enhance these vulnerabilities, and energy security is likely to remain focused on preventing disruptions.

South Asia

Although South Asia averages very low levels of per capita consumption, it accounted for 3.9 percent of world energy consumption in 2000, up from 2.8 percent in 1991. The region is an emerging demand market, which has some advantages of proximal location with respect to supply sources of the Middle East and Central Asia.

Due to the significant variation – in commercial energy mix, energy expectations of countries of the region are very different.

The dominant source of energy mix among countries of the region is as follows:

(a) Bangladesh   –   Natural Gas (68%)
(b) Bhutan   –   Hydroelectric
(c) India   –   Coal (53%)
(d) Maldives   –   Petroleum (Near 100%)
(e) Nepal   –   Hydroelectric
(f) Pakistan·   –   Natural Gas (45%)
(g) Sri Lanka   –   Petroleum (78%)

Demand is expected to rise mainly for oil, gas and coal. Despite measures taken towards diversification in respect of oil and gas supplies, the heavy dependence on the Persian Gulf is likely to continue for the foreseeable future. There are high expectations from the new sources of Caspian Sea/Central Asia. Countries bordering the Persian Gulf region have also articulated "grand visions" of becoming "Central Asia's Gateway to Emerging Markets." Numerous practical difficulties have, however, acted as a constant "dampener" to heightened expectations. In addition, there are expectations to build new "strategic relationships" based on a convergence of energy interests. Expectations are also high from the exploration and production (in own areas as well as overseas), and from major powers (global and regional) to maintain strategic stability. Some countries also have major expectations from nuclear energy.

Sub-regional collaboration based on shared energy security interests has been marginal. Perceptions of vulnerabilities are heavily focused on the security of pipelines of the future and the security of SLOCs. Vulnerabilities are also seen in terms of political and economic obstacles like developments in Afghanistan, elusive cooperation on the pipelines issue between India and Pakistan and an increase in low-intensity threats at sea. Some countries have enhanced vulnerabilities simply by the pattern of SLOCs with respect to geography, which renders the entire war-fighting capability susceptible to disruptions in oil and gas supplies. Rising levels of consumption of gas are likely to create additional vulnerabilities, since the security of sea lanes as well as pipelines coming from diverse sources would become increasingly challenging. There is also a heightened sense of vulnerability related to disruptions by sabotage or acts of terrorism.

There are more voices in the region highlighting rapidly increasing levels of emissions, effluents, solid waste and nuclear safety. Economic development versus the cost of environmental safeguards is a relatively less-explored area and one which is likely to gain significance in the years ahead. Challenges with respect to the use of coal include opencast mining, reserve depletion, ash generation, high cost of transportation, and improving efficiencies at coalmines.

South East Asia

Energy demands in South East Asia are set to rise consequent on the recovery from the financial crisis, a gradual move towards manufacturing and an increased emphasis on infrastructure. Despite oil producers like Indonesia, Malaysia, Thailand and Vietnam, energy self-sufficiency seems distant. Only Indonesia is a significant oil and gas exporter, though its continued role in that capacity is somewhat uncertain. Oil production has been declining, although gas exports are likely to increase in the light of new offshore gas discoveries in the Natuna field in the South China Sea and the Wiriage Deep field off the coast of New Guinea.

Countries of the region have major expectations from a shift to gas as the primary source, which, in turn, would enhance self-sufficiency and reduce dependence on the Middle East. Sub-regional cooperation based on shared energy security interests, particularly for developing indigenous resources, has further raised expectations related to self-sufficiency. Projects to increase the flow of oil through pipelines, in order to relieve maritime congestion, are also in progress. Since the region sits astride the strategic chokepoint of the Malacca Straits, there are expectations of cooperation, including from non-ASEAN countries, with respect to SLOC security.

Within the overall framework of ASEAN, the coordination of energy-related policies is seen as a major guarantee for energy security. These include an ASEAN Council on Petroleum (ASCOPE), the Heads of ASEAN Power Utilities/Authorities (HAPUA), the ASEAN Ministers on Energy Meetings (AMEN), and the ASEAN Energy Business Forum (AEBF). Despite many institutional arrangements, observers feel that concrete collaboration on energy-related policies has not yet been substantial.

Energy-related vulnerabilities are seen in terms of outstanding bilateral and territorial disputes, sovereignty issues in respect of disputed islands, the choke points of the Malacca straits and SLOCs, piracy and illegal trafficking at sea, and heavy dependence on trade and commerce. The rise of terrorism in the region, increasing urbanization and the impact of energy security strategies of neighbouring countries are other areas of concern.

East Asia

Many countries of East Asia, though not part of the IOR, have considerable significance when examining energy security in the IOR. The three leading oil consumers of Asia are in this region (Japan, China and South Korea), and depend heavily on the trade routes of the Indian Ocean for their supplies.

From the Maoist era vision of energy self-sufficiency, China has rapidly adapted to the new realities of increasing imports of oil. After nearly 30 years of being an exporter, China became a net importer of oil in 1993. While coal is the dominant energy source, meeting about 75 percent of domestic energy demand, a significant growth in demand of oil and gas is forecast. Gas demand is expected to grow at around nine percent in the next ten years, and dependence on the Middle East for oil is expected to increase to about 75 percent of total oil consumption. Growth in the nuclear sector has been slow, particularly after nuclear power plant construction was halted for some time.

China seeks to ensure its energy security for the future by developing new energy linkages, particularly with its neighbours. There are major expectations from penetration in "niche markets." Equity purchases of foreign oil and long-term supply agreements from such equity positions are expected to strengthen energy security (these include Angola, Canada, Indonesia, Kazakhstan, Kuwait, Mongolia, Nigeria, Papua New Guinea, Peru and Saudi Arabia). Oil from distant markets would be obtained on a "swap" basis, from the Persian Gulf by sea and from Kazakhstan by the Kazhak- Xinjiang pipeline. Expectations from Central Asia, for oil as well as gas, are also linked with expectations of increased convergence with Russia, including from Irkutsk in the Russian Far East. Increased expectations from gas have also led to more emphasis on upgrading the internal distribution infrastructure.

Expectations are also based on using the environmental impact of high polluting coal as leverage. Many strategists in China favour strengthening intra-regional cooperation and have often proposed a regional energy organization for East Asia. It is also suggested that the use of advanced technologies, including renewables and coal-gasification, can help mitigate Chinese vulnerabilities while meeting development needs.

Perceptions of vulnerabilities range from long trade routes and choke points in the IOR, geopolitics related to proposed pipelines, financial aspects and dependence on markets, the role of extra-regional powers in the IOR, a lack of infrastructure, the introduction of advanced technologies, and the limitations of hydropower and nuclear power.

Japan remains heavily dependent on foreign supplies for energy. Despite increases in nuclear energy use and gas (about 12 percent – Japan is the largest importer of LNG), oil continues to be its primary source (at about 51 percent). On its own, Japan is poorly endowed with natural resources.

Japan has expectations from gas deposits in the Russian Far East and Siberia. Proposals include pipeline loops from Sakhalin to Hokkaidu, and from Irkutsk via China. Engagement with China on energy issues has raised expectations on the larger question of overall security. The Central Asia/Caspian Sea region is also expected to ameliorate heavy dependence on the Middle East, notwithstanding long distances that would be involved in the supply network. There are expectations from the ongoing security relationships and convergence of interests with countries like South Korea and India. There are also expectations from a more proactive engagement with the Middle East region.

Being an island state, vulnerabilities are seen mainly in terms of shipping lanes, particularly in view of the heavy dependence on the Middle East. Some writers, based on a scenario of an energy-hungry neighbour becoming increasing aggressive, also articulate a perception of vulnerability. Environmental vulnerabilities include the impact of Chinese coal usage (The Japan-China environmental cooperation programme seeks to mitigate such vulnerabilities through financial aid from Japan). Concerns about safety related to nuclear installations are a dominant theme, bringing in some uncertainty about the future expansion of the nuclear sector.

South Korea has virtually no proven oil or gas resources and depends almost entirely on imports. Three quarters of its oil imports come from the Persian Gulf region and most of its LNG imports come from South East Asia.

Like Japan, South Korea has expectations from Russia's gas reserves. Diversification of supplies is expected to ease the dependence on the Middle East. In addition, there are expectations from overseas exploration, increased gas usage and expansion of the nuclear sector. Perceptions of vulnerability include shipping lanes from the Persian Gulf through the IOR, rising industrial demand, concerns about energy efficiency and susceptibility to market fluctuations. Lack of any proven indigenous reserves heightens the sense of vulnerability and weak bargaining position. While there is some recognition of convergence of interests with Japan, concrete coordination driven by energy security considerations is yet to take place.

Conclusion

The Indian Ocean Region is key to the emerging geopolitics of energy security. Yet, a comprehensive look at the subject requires an analytical and conceptual framework that can be used as a template for the purpose. The broad contours of energy security need to be outlined in a framework that captures diverse, sometimes competing, views, emphasis, orientations and postures. Security concerns based on energy as the driver also vary among countries and organizations, and energy security strategies conceived depend largely upon perceptions and understanding of energy expectations, energy interdependence and energy vulnerabilities. Based on such overall assessments, integrated strategies involve mitigation as well as defensive measures at the sub-regional, national and sub-national levels.

In the IOR, there is considerable variance in expectations of different suppliers as well as demand markets. Notions of energy competition also differ significantly, although most countries are focused on maintaining reliable flows and stability in prices over the long-term. Oil and gas continue to be at the forefront of the debate, and are likely to remain so for the foreseeable future. Coal, nuclear energy and hydropower have significant relevance, too, although potential competition over prospective pipeline projects and the security of sea lanes are the main pillars of emerging strategies. While many countries of the region do not have an explicit linkage between energy security and military strategy, a trend in that direction is clearly discernible. Issues related to environmental security, too, have gained increased recognition in recent years, making development of a coherent strategy even more challenging.

New energy security concerns have added to some traditional ones and numerous challenges can be foreseen over the next 15 to 20 years. The thin line between competition and cooperation based on such concerns may be getting thinner, and a better understanding of how others conceptualize energy security of the future can swing the odds in favour of greater cooperation.

References

Ardebili, H. K. (2003), "Geopolitics of Energy Security and I.R. Iran's Position," paper presented the 8th Annual Conference of Institute For International Energy Studies, Teheran, 29-30 November.

Barnett, T. P. M. (2002), "Asia: The Military Market Link," U.S. Naval Institute Proceedings (January), pp. 53-56.

Batra, R. K. (2004), "Will Coal Remain King," *Business India*, 5-18 January 2004, 108-110.

Billings, R. (2004), "Global Energy Outlook," presentation made at the 5th Indian Oil and Gas Conference, New Delhi, India, 12-14 January 2004.

BP Amoco (2003), Statistical Review of World Energy, at <www.bpamoco. com > [30 December].

Cordesman, A. H. (1999), "Geopolitics and Energy in the Middle East", 15 September, <http://www.csis.org/mideast/reports/Menergy. html > [15 January 2004].

DeLaquil, P., Chen Wenying, Eric D. Larson (2003), "Modeling China's Energy Future," *Energy For Sustainable Development*, Volume VII, Number 4, December, pp. 40-56.

Guoxing, J. (1998), "Energy Security: A View from China," *Korean Journal of Defence Analysis*, (Winter).

Hermsmeyer, G. A. "Oil, Security and the Post-9/11 World," Jerome E. Levy Occasional Paper No. 4, Economic Geography and World Order, U.S. Naval War College, 21-38.

India Tata Energy Research Institute (TERI) (2002), Defining an Integrated

Energy Strategy for India, TERI, New Delhi.

Kang Wu and Fereidun Fesharaki, (2002), "Managing Asia Pacific's Energy Dependence on the Middle East: Is There a Role for Central Asia?", U.S. East West Centre Asia Pacific Issues Analysis, No. 60, June.

Kyodo News (1998), MITI Panel Debates Revising Long Term Energy Outlook, cited in Foreign Broadcasting Information Service, 26 January.

Liotta, P. H. (2002), "Boomerang Effect: The Convergence of National and Human Security," Security Dialogue, *Peace Research Institute of Oslo*, Volume 33, Number 4, December, pp. 473-488.

Luthra, G. (2002), "Energy Security: Scenarios for the Asian Region," paper presented during seminar at the United Service Institution of India, New Delhi, 12 September.

Pachauri, R. K. (2004), "Not All Hot Air: Indo-Pak Peace in the Pipeline," *The Times of India*, 05 February 2004.

Ratnam, C. (2002), "Safeguarding of India's Energy Security," USI National Security Paper (P20), USI National Security Series, United Service Institution of India, New Delhi.

Sharma, S. (1991), "Structural Change and Energy Policy in ASEAN," *Energy Market and Policies in ASEAN*, eds. Shankar Sharma and Fereidun Fesharaki (Singapore: Institute of Southeast Asian Studies), pp. 57-80.

Sikri, R. (2002), "India's Energy Security: Some Foreign Policy Aspects," Unpublished Paper, March.

Smil, V. (1998), "China's Energy Resources: Continuity and Change," *China Quarterly*, (December).

Stewart, A. (1998), "Russia and China's Interests in the Middle East and Caspian," *Geopolitics of Energy*, (April).

U. S. Asia Pacific Centre For Security Studies (1999), Energy Security in the Asia Pacific: Competition or Cooperation?, Seminar Report, 15 January, at <http://www.apcss.org/Publications/Report_Energy_Security_99.html > [02 January 2004].

U.S. Centre For Strategic and International Studies (CSIS) (1998), The Changing Geopolitics of Energy, 12 August, at <http://www.csis.org > [20 December 2003].

U.S. Energy Information Administration (EIA), Regional and Country Briefs, at <www.eia.doe.gov > [25-25 December 2003].

U.S. Institute For Advanced Strategic and Political Studies, Caspian Great Game: The Current State of Play, at <http://1w11fd.1aw11.hotmail.msn.com/cgi-bin/getmsg?curmbox=F000000005&a=860dbc6...> [23 May 2003].

CHAPTER 3

The Geopolitics of Global and Indian Ocean Energy Security

Dennis Rumley

Introduction

This Chapter is part of a larger project aimed at developing an integrated systems approach to the geopolitics of energy security (Table 3.1). It is concerned principally with elaborating some of the basic dimensions of energy security both globally and more specifically in the Indian Ocean region and briefly outlines some of the policy responses undertaken by states in relation to questions of energy security. As has been argued, there exist at least three facets to the maximisation of a state's energy security – limiting vulnerability to energy dependence on another state or region, providing long-term supplies at reasonable cost, and, third, ensuring that the international energy system functions in an ecologically sustainable manner (Martin, Imai and Steeg, 1996, 4). Over the past thirty years, global energy consumption has almost doubled, and, while national economies use significantly less energy per dollar of economic output compared with 1970, by 2030 global energy demand is still likely to increase by a further 50 percent (Kelly, 2004). In this environment, and coupled with yet another oil shock in 2004, energy security questions take on a new urgency in the 21st century and the nature of policy responses becomes even more critical. The Chapter is divided into five broad sections – first, the nature of global and regional geopolitical change and energy insecurity will be considered; second, there

Table 3.1. A simplified systems framework for energy security.

1. Dynamics of Energy Demand and Supply Relations

(i) Historical patterns of energy use

Economic growth and demand
The energy consumers
21st C shift towards the developing world and especially towards Asia
Competition for energy – potential for resource conflicts

(ii) Supplies: Dependency

Dependency on imports for a significant proportion of energy needs –
    concept of an energy-dependent state
Dependence on a single state or region for energy supplies
Dependency on a single source of energy
Structural energy dependency
Energy vulnerability and energy mix → nuclear

(iii) Source: Resource holder

Energy exports and reserves
Concept of an energy-niche economy
The global energy market – the role of energy-producing/consuming
    groups – for example, OPEC, IEA, MNCs
Strategic significance of resource holders – for example, Iran, Russia
"Oil as a weapon"
Stability of resource holders – for example, Iraq, Saudi Arabia
Resource Diplomacy – for example, Australia, Japan, USA
Future resource holders – for example, EEZs – competition for potential
    sources?

2. Energy Flows

Energy type and transportation prospects – coal by land and sea; gas by
    pipeline or by liquefaction; oil by tanker; the uranium trade
The transport holder
Infrastructure bottlenecks
Strategic implications of energy flows/routes
Terrorism, naval interference, congestion and environmental problems

3. Environmental Outcomes

Variations of emissions by energy type
Environmental implications of each energy source
"Hydrocarbon Man" and the "Hydrocarbon Society"
Environmental impacts of energy exploration
Trade off between environmental impact and energy needs?

International agreements to reduce emissions and adherence to these

4. *Policy Responses to Enhance Energy Security*

(i)	Spatial	Concept of an energy foreign policy
		Hemispheric policies – for example, USA
		Access new sources of supply – Atlantic Basin, Caspian, Russian, Far East
		Import reduction strategies
		Cooperative energy security – energy community concept
		Maintain open shipping routes
(ii)	Energy Transfer	Global energy scenarios
		Diversification – enhances energy security
		Greater exploration – capital and technology holder
		Stockpiling/reserves
		Flexible switching
		Alternative renewable energy sources – wind, wave, hydrogen, etc
(iii)	Non-spatial	Management of dependency – energy independence?
		Let the market rule
(iv)	Environmental	Energy efficiency
		Sustainable development policies – towards decarbonization

will be a brief overview of the structure of global and regional energy consumption; third, the issue of state energy dependence will be elaborated; fourth, there will be an evaluation of the geopolitics of resource holders, and, finally, some policy responses to questions of energy insecurity will be reviewed. Most discussions about energy security to date have been grounded in a realist framework emphasising traditional views of security. Among other things, the outcome of this is that "'Energy security' is seen as the practice of 'securing' energy for the nation-state" (Barnett, 1997, 442). It will be argued here that, while concern over energy security will likely increase the probability of inter-state conflict, the development and maintenance of a truly secure global and regional future requires an increase in inter-state cooperation in all aspects of energy production and use, as well as a much greater collective concern for the global environmental impacts of energy consumption.

Geopolitical Change and Energy Insecurity

It has been argued that, from a Western perspective, the nature of global geopolitical and geoeconomic change necessitates that at least five main factors need to be taken into account in the adoption of a new geopolitical paradigm – "a geography of strategic interactions". These are that Europe is now more secure than ever and is no longer a significant military power; second, that there exists a multipolar "balance of terror" which stretches in an unbroken belt from Israel to North Korea and within which there are few Western allies; third, that this "arc of terror" cuts across the military and political theatres into which the West divided Asia; fourth, that the great interior of Asia is "in play" again, and, fifth, that the world energy map is being redrawn as many Asian states industrialise and grow and Asian energy reserves await exploration and exploitation (Bracken, 2000, 2-3).

Issues of energy security are clearly going to be of increasing Asian as well as global importance associated both with increasing competition for resources and with the prospect of the exploitation of new resources resulting in changes to the global flows of energy. In both cases, states around the Indian Ocean take on an even greater geopolitical significance, not only due to their proximity to ocean energy routes, but also because of their significance either as suppliers of or destinations for increasingly scarce energy resources.

Energy scarcity and the geography of energy resource endowment will not only lead to greater competition for the control over resources, but it will also increase the potential for interstate conflict:

> The world's voracious appetite for energy is likely to fuel competition for diminishing supplies of fossil fuels and heighten national anxieties in energy-deficient states about long-term energy security (Dupont, 2001, 70).

> Such accounts of energy security are common; they countenance the projection of force to secure uninterrupted supplies of oil, and the pragmatic and at times forceful engagement with nations who by dint of (mis)fortune have energy resources but apparently little sovereignty (Barnett, 2001, 35).

As prices rise, moreover, contending groups and elites in resource-producing countries will have greater incentive to seize and retain control of valuable mines, oil fields and timber stands. The result, inevitably, will be increasing conflict over critical materials (Klare, 2002, 20-1).

Access to new sources of energy in "The Eurasian Balkans" will likely "stir national ambitions, motivate corporate interests, rekindle historical claims, revive imperial aspirations, and fuel international rivalries" (Brzezinski, 1997, 125).

On the one hand, in order to maximise economic and political stability, it has been suggested that it is essential that the West pursue policies which encourage "geopolitical pluralism", especially in regions containing new sources of energy such as in the so-called "Eurasian Balkans". From a United States perspective, this would mean helping to "ensure that no single power comes to control this geopolitical space and that the global community has unhindered financial and economic access to it" (Brzezinski, 1997, 148). Geopolitical and geoeconomic control over energy-rich regions and energy routes is likely to become increasingly contested in an environment of scarcity and greater competition.

The Structure of Global and Regional Energy Consumption

During this century, apart from continued strong demand in the industrialised world, the major growth in energy demand will be in the developing world, and especially in Asia (BPSRWE, 2003). Competition for energy among these states will become increasingly intense, particularly among the largest consumers as well as among states in the northern rim of the Indian Ocean and in Northeast Asia. This is especially problematical for those states with small and dwindling energy resources (Dupont, 2001, 70).

Due principally to reasons associated with a combination of economic security, energy security and environmental security, the structure of energy use has changed somewhat in the latter part of the 20th century. While oil remained dominant, there was a shift away from oil and coal towards gas and nuclear (Figure 3.1).

Not surprisingly, the consumption of energy by volume and type varies significantly by state and by level of economic development. The ten largest global consumers of oil, natural gas, coal and nuclear energy use 59.9%, 65.1%, 80% and 85.8% of the global total respectively. Furthermore, clearly, given the size and nature of its economy, the United States is the largest consumer in all but one of these categories (Table 3.2).

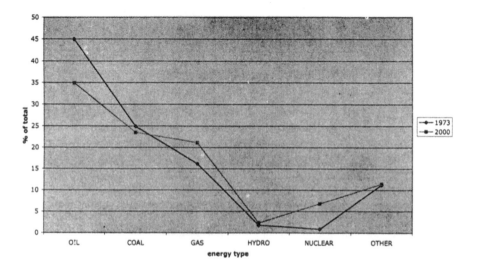

Figure 3.1. Global primary energy supply 1973-2000.

In the Indian Ocean Region (IOR), while India is an important global consumer, especially of coal (7.5% of the world's total) and oil (2.8% of the world's total), Iran and Saudi Arabia are important global consumers of natural gas, and South Africa and Australia also consume globally significant volumes of coal (Table 3.2). From a regional perspective, there is considerable variation in terms of energy mix among IOR-ARC states (Table 3.3). While the regional importance of hydro is small and of nuclear is almost non-existent, the energy mix is dominated by varying combinations of oil, natural gas and coal. If we take 'energy dominance' to equate with a minimum of 50 per cent of primary consumption from one energy category, then we can say that, for Indian Ocean states, Indonesia, Singapore and Thailand are oil-dominant, Bangladesh, Iran and UAE are natural gas-dominant and India and South Africa are coal-dominant. Australia (coal and oil) and Malaysia (natural gas and oil), on the other hand, possess a kind of 'dual energy dominance'.

Table 3.2. Global energy consumption – the world's 10 largest consumers (E-10) (as % of world total).

oil	natural gas	coal	nuclear
USA 25.4	USA 26.3	China 27.7	USA 30.5
China 7.0	Russia 15.3	USA 23.1	France 16.2
Japan 6.9	UK 3.7	India 7.5	Japan 11.7
Germany 3.6	France 3.3	Japan 4.4	Germany 6.1
Russia 3.5	Canada 3.2	Russia 4.1	Russia 5.3
S Korea 3.0	Japan 3.1	Germany 3.5	S. Korea 4.4
India 2.8	Ukraine 2.8	S Africa 3.4	UK 3.3
France 2.6	Iran 2.7	Poland 2.4	Ukraine 2.9
Italy 2.6	Italy 2.5	Australia 2.1	Canada 2.8
Canada 2.5	S Arabia 2.2	S Korea 2.0	Sweden 2.6

Source: BP Statistical Review of World Energy 2003

Table 3.3. IOR-ARC energy mix 2002.
(% primary energy consumption)

	Oil	Natural Gas	Coal	Nuclear	Hydro
Australia	33.7	19.1	43.8	0	3.4
Bangladesh	24.1	71.6	2.8	0	1.4
India	30.1	7.8	55.6	1.4	5.2
Indonesia	50.0	30.6	17.4	0	2.1
Iran	45.8	52.6	0.7	0	0.9
Malaysia	43.4	46.9	6.4	0	3.3
Singapore	95.7	4.3	0	0	0
South Africa	21.6	0	74.9	2.7	0.8
Thailand	51.2	33.8	12.5	0	2.3
UAE	25.9	74.1	0	0	0

Source: BP Statistical Review of World Energy 2003.

Indian Ocean states with more than 70 per cent consumption from one category can also be regarded as 'type dependent'. Thus, Singapore is dependent on oil as an energy source, Bangladesh and UAE are dependent on natural gas and South Africa is dependent on coal (Table 3.3). While a situation of energy dominance is one of potential uncertainty, energy-type dependence can clearly result in increased vulnerability to supplies and to market vicissitudes. This situation is exacerbated in contexts of import dependence.

The Problem of Energy Dependency

If we take an arbitrary cut-off point for net imports of total energy of greater than 45 million tonnes of oil equivalent (mtoe), then it is possible to identify thirteen global 'energy-import dependent' states, the majority of which import far more than domestic energy production and several of which are among the world's largest energy users (Table 3.4).

Table 3.4. Energy-import dependent states 2001.
(net imports >45 mtoe)

	Net Imports	Domestic Production (DP)	Imports as % of DP
1. USA*	641.7	1711.8	37.5
2. Japan*	417.1	104.1	401.2
3. Germany*	217.2	133.7	162.5
4. South Korea*	164.4	34.2	480.7
5. Italy	146.7	26.3	557.8
6. France*	133.5	133.2	100.2
7. Spain	100.6	33.0	304.8
8. #India*	91.0	438.1	20.8
9. Taiwan	82.3	10.8	762.0
10. Ukraine	58.2	83.4	69.8
11. Belgium	52.1	13.1	397.7
12. #Singapore	47.5	0.1	47500
13. Turkey	46.0	26.2	175.6

IOR-ARC member;   *E-10 member
Source: Key World Energy Statistics, IEA 2003.

Two of these states – India and Singapore – are IOR-ARC members, and, clearly, given the size of India's domestic production, Singapore is much more energy-import dependent. Furthermore, while India is much more reliant on coal imports, Singapore possesses a very high level of oil-import dependency. If we analyse the nature of energy-import dependency for all IOR-ARC states, we find that, apart from India and Singapore, Thailand also has an important degree of dependency. On the other hand, several Indian Ocean states are net exporters, especially Australia, Iran and UAE (Table 3.5).

Table 3.5. IOR-ARC: Net energy imports 2001.*
(mtoe)

net importers		net exporters	
India	91.0	Australia	-134.1
Singapore	47.5	Iran	-126.0
Thailand	36.0	UAE	-105.2
Bangladesh	4.3	Indonesia	-80.9
Sri Lanka	3.6	Oman	-55.8
Kenya	2.8	South Africa	-32.8
Tanzania	0.9	Malaysia	-25.7
Mozambique	0.2	Yemen	-19.0

*data unavailable for Madagascar and Mauritius.
Source: Key World Energy Statistics, IEA 2003.

Since oil still remains the most important primary energy source, then the nature of oil import dependency takes on a special geopolitical significance. If we examine the world's four largest oil importers – Europe, USA, Japan and China – varying degrees of 'regional oil dependency' can be identified. Thus, Europe relies more for its oil on the former Soviet Union (but has other major sources both from the Middle East and Africa); the United States 'hemispheric' policy is dominant; Japan is dependent upon the Middle East; China's largest supplies are also from the Middle East but it has other important Asia-Pacific supplies (Table 3.6). As has been noted, moving from a single supplier to a small number of regional suppliers reduces the effect of dependency to one of vulnerability (Bahgat, 2001, 526).

Energy insecurity can become especially acute as a result of what has been referred to as "structural energy dependency". For example, Ukraine's energy dependency (see Table 3.4) is partly a product of inflexible structural ties inherited from the Soviet period which effectively inhibit switching to other suppliers (Balmaceda, 1998, 258). For many states (like Ukraine), energy vulnerability makes nuclear power very attractive (Calder, 1996, 62), and, in terms of nuclear shares of national electricity generation, two states are 'nuclear dependent' – Lithuania (78%) and France (77%) – while another two - Belgium (58%) and Slovakia (53%) – are 'nuclear dominant' (IAEA, 2000). In the Indian Ocean Region, however, only three states currently generate electricity from nuclear power, and the amounts are quite small – South Africa (7%), India (4%) and Pakistan (3%).

Table 3.6. World's largest oil importers and their suppliers 2002.

% of world's imports	% imports from				
	Middle East	Former USSR	N&W Africa	W Hemis*	Other A.P.
Europe 27.3	27.4	36.5	20.9	4.0	0.8
USA 26.1	20.4	1.7	12.3	51.9	1.5
Japan 11.6	78.0	0.5	3.0	0.4	11.3
China 4.7	38.7	8.1	9.8	0.9	28.3
World	41.6	12.3	13.2	15.8	4.8

*Canada, Mexico, C and S America
Source: BP Statistical Review of World Energy, 2003, page 18

The Geopolitics of Resource Holders

Apart from dependency, energy security is also a function not only of resource endowment, but also as a result of access to transport and to capital and technology. In this regard, a useful distinction has been made between "resource holders", "transport holders" and "capital and technology holders". Resource holders include those states which possess significant energy resources but must rely on others for transport to market and for investment capital – for example, Turkmenistan. States which occupy territory required for transit to market are seen as transport holders – for example, Pakistan, in the case of the proposed "peace pipeline". Capital and technology holders are those actors which are able to offer capital and technology but are neither resource nor transport holders – for example, Japan (Cutler, 1999).

From the point of view of resource holders, clearly, any analysis of energy security needs to take into account the role of energy corporations (ECs) as well as the role of other public and private energy producing and consuming groups such as OPEC and IEA. The ECs are likely to be of three principal types - privately-owned multinational corporations operating from states with considerable energy reserves (for example, those based in the United States); state-operated corporations functioning in states with large reserves (for example, in Iran); and, third, corporations which are operating on behalf of states which are energy-deficient (for example, those in Japan).

Table 3.7. Global energy reserves 2002.
(% of global total)

oil	natural gas	coal	uranium
Saudi Arabia (25%)	*Russia (30.5%)	*USA (25.4%)	Australia# (27.8%)
Iraq (10.7%)	Iran# (14.8%)	*Russia (15.9%)	Kazakhstan (15.2%)
UAE# (9.3%)	Qatar (9.2%)	*China (11.6%)	*Canada (14.1%)
Kuwait (9.2%)	Saudi Arabia (4.1%)	*India# (8.6%)	S Africa# (9.6%)
Iran# (8.6%)	UAE# (3.9%)	Australia# (8.3%)	Namibia (7.6%)
Venezuela (7.4%)	*USA (3.3%)	*Germany (6.7%)	Brazil (6.3%)
[Australia# (0.3%)]	[Australia# (1.6%)]	S Africa# (5.0%)	*Russia (4.2%)

#IOR-ARC member; *E-10 member
Sources: BP Statistical Review of World Energy 2003; World Nuclear Association 2002.

Most states outside of the United States with significant global energy reserves would qualify as resource holders. Furthermore, in conditions of potential scarcity, they take on a new global geopolitical significance. In the case of IOR-ARC, this especially means UAE and Iran in terms of oil and natural gas, India, Australia and South Africa in terms of coal, and Australia and South Africa in terms of uranium (Table 3.7).

From a global perspective, for resource holders there is often a very great difference between national energy reserves, on the one hand, and energy consumption, on the other hand. Net energy imports therefore provide a preliminary means of identifying resource holders. Indeed, some resource holders may be better categorised as energy-niche economies since they are both considerable net energy exporters and are economies that are economically dependent on such exports.

Globally, it is possible to identify at least seventeen energy-niche economies, and, of these, five are IOR-ARC members – Australia, Iran, UAE, Indonesia and Oman – and a further eleven are members of OPEC. Of these economies, only Canada, Mexico, Norway and Russia are neither IOR-ARC nor OPEC members and only three states are members of both groups – Indonesia, Iran and UAE (Table 3.8).

Table 3.8. Energy-niche economies 2001.
(net imports –45mtoe or more)

1. Russia	–367.3
2. Saudi Arabia*	–364.2
3. Norway	–201.3
4. Venezuela*	–155.8
#5. Australia	–134.1
6. Canada	–131.6
#7. Iran*	–126.0
8. Algeria*	–115.5
9. Nigeria*	–110.3
#10. UAE*	–105.2
11. Iraq*	–94.8
12. Kuwait*	–92.0
#13. Indonesia*	–80.9
14. Mexico	–76.8
15. Libya*	–58.3
#16. Oman	–55.8
17. Qatar*	–47.8

IOR-ARC member   *OPEC member
Source: Key World Energy Statistics, IEA 2003

The economic and political stability of such resource holders, while critical to the regional and global economy, can also be seen to be incorporated into a state 'resources diploma' y' strategy –

that is, the global 'marketing' of the state as a stable and reliable long-term energy supplier. However, future resource holders in the trans-Caspian, in Central Asia and beneath the Asian continental shelf will likely significantly affect the direction of oil and gas flows around the globe (Bracken, 2000, 3).

Some Policy Responses to Enhance Energy Security

There are at least four broad categories of state policy responses to the problems raised by energy security which will be raised here – spatial/geopolitical, energy transfer, non-spatial and environmental – and each of these involves a range of possibilities (Table 3.1). An additional, and potentially even more important policy response, which relies on interstate cooperation is the energy security community concept.

Spatial/Geopolitical

One of the principal goals of the US National Energy Policy is to "increase energy security" which is seen as a priority for US trade and foreign policy. Part of this strategy is the need to build strong relations with energy-producing states around the IOR, but particularly with those located in its hemisphere (NEPDG, 2001). As is well known, the search for scarce resources is a traditional focus of American foreign policy. In the case of the Eurasian Balkans strategy noted earlier, a US policy of "geopolitical pluralism" would likely ensure American dominance. Indeed, from a global perspective, one of the several explanatory reasons which has been put forward for the more than 725 US military bases around the world is to provide protection for American dependence on imported energy supplies (Johnson, 2004, 167).

Maintaining proximate supplies is an important additional strategy, both as a result of the provisions of NAFTA and also of the Hemispheric Energy Initiative (HEI) which encourages Canada, Mexico, Venezuela and other hemispheric states to maintain strong energy linkages with the United States (NEPDG, 2001, 8-10). HEI is thus a kind of energy community concept akin to energy regionalism and is responsible for more than 50 per cent of US oil imports (Table 3.6).

The hemispheric approach is also in part a response to concerns over political instability within the Middle East. Similar concerns have fuelled the search for new sources of energy in the Atlantic Basin, the Caspian and the Russian Far East (Fesharaki, 1999, 93). Undeveloped energy reserves in the IOR will continue to attract the attention of large energy-consuming states. In Africa, for example, while most export interest has centred on West and North African energy deposits (NEPDG, 2001, 8-11), increasing exploration is taking place on the East African coast, which, at present, is totally energy-dependent (Wilkinson, 2003).

Three necessary components have been suggested for a bolder strategy of "cooperative energy security" advocated for the Caspian Region – an investment-friendly environment, guarantees of secure transport and political stability (Cutler, 1999, 253). Of course, given the importance of the Indian Ocean as an energy routeway, clearly the maintenance of secure and open shipping routes is a pre-requisite for regional as well as global energy security (Fesharaki, 1999, 91). This is likely to become increasingly important as supply chains lengthen.

Energy Transfer

Various 'energy transfer' policies are available to energy-deficient states. In particular, diversity of supply region and diversity of energy type are key strategies that are seen to increase energy security (NEPDG, 2001, 8-6). Greater exploration for increased supplies, on the other hand, is inevitably in the hands of capital and technology holders. The maintenance of strategic stockpiles has been seen as one strategy for maximising energy security. Indeed, each of the IEA members is required to hold stocks equivalent to 90 days or more of its net imports (NEPDG, 2001, 8-16). An ability to switch from one energy source to another is also potentially advantageous. Perhaps most importantly of all is the search for alternative renewable energy sources – wind, wave, hydrogen, fusion, etc. Import reduction strategies both at home and abroad have been advocated in order to increase domestic energy security. One commentator went so far as to suggest that "keeping China and Indonesia from becoming large-scale energy importers likewise deserves priority" (Calder, 1996, 65).

While Northeast Asian states currently account for more than one-third of global coal consumption, environmental considerations will mean that increasing demand pressure will be placed on oil, natural gas and nuclear energy. However, there is likely to be considerable variation by state in terms of the attractiveness of alternatives to coal. In the case of nuclear energy, for example, Japan and Taiwan's nuclear capacity is likely to increase more slowly than other regional states. Both Japan and Taiwan currently obtain about one-third of electricity from nuclear power. In contrast, South Korea, with approximately the same proportion of its electricity generated from nuclear power is set to double over the next two decades. Furthermore, China, with about 1% of its electricity from nuclear power is set to increase dramatically in percentage terms to 2020 (US Energy Information Administration). This will therefore place considerable pressure on the further development of Indian Ocean state uranium reserves, especially those in Australia (Table 3.7).

Non-Spatial

For energy-import dependent states, debates over the need to achieve energy self-sufficiency are likely to be long and potentially fruitless. It has been argued, for example, that, in the case of the United States, oil-import dependency is set to continue indefinitely and thus a more realistic approach to energy insecurity is to endeavour to identify new ways of managing energy dependence (Bahgat, 2001, 542). These mechanisms may well involve a combination of various elements of policy responses outlined here.

One important and contested 'non-spatial' neo-liberal energy security strategy is for states to rely completely on the marketplace. One of the central propositions of this view is that the operation of global markets will overcome most traditional energy security concerns. Indeed, some would go so far as to argue that the market is a "source of energy security" (Fesharaki, 1999, 88). Such an optimistic view of the market is essentially predicated on the belief that stability in energy prices and security of supplies is of overriding benefit to both producers and consumers (Bahgat, 2001, 542). Producers are thus keen to maintain stability in price and supply in order to ensure their own economic security. In reality, however, from time to time the geopolitics and geoeconomics of energy security are likely to distort this notion of interdependency, induce occasional "oil shocks" and lead to greater competitive and potentially conflictual pressures. The political economy of energy markets is therefore much more likely to encourage the largest consumers to intervene in the politics of resource holders in an environment of increasing scarcity.

Environmental

While state competition for access to other resource-rich states and regions is closely linked to geopolitical competition and control and associated potential conflict, it has been argued that the environmental security implications of increased energy use could well be of even greater longer-term significance (Dupont, 2001, 87-8). This is primarily as a result of essential energy requirements being met by non-renewable sources all of which have varying degrees of negative impact upon the environment and hence upon public health. Each energy source possesses its own economic, environmental and health benefits, costs and risks. Clearly, maximising energy security also involves fundamental questions related to human and environmental security. Choosing any particular energy strategy inevitably means choosing an environmental strategy and a fundamental policy shift towards renewable energy is only on the distant horizon.

As has been argued, the global implications of this are especially problematical since a consensus on a long-term plan for a sustainable and safe energy future has yet to be reached (Brundtland, 1990, 212-249). Reaching such a consensus, however, is likely to generate new tensions and potential conflicts. On the one hand, the global shift in recent decades away from oil and coal and towards gas and nuclear as primary energy sources (Figure 3.1) is to be welcomed from the viewpoint of fossil fuel emissions. On the other hand, it raises many other important issues, not the least of which relates to nuclear safety and to the handling and national, regional and global disposal of nuclear waste. In addition, it raises potential conflicts associated with the actual mining and export of uranium. As far as the Indian Ocean Region is concerned, it has not only been suggested that Western

Australia might well be an ideal site not only for national nuclear waste disposal but that the opening of its first uranium mine is inevitable in the face of growing international demand (Williams, 2005).

In the longer-term, policy pressures towards 'decarbonisation' will negatively impact upon Indian Ocean coal producers especially in Australia, India and South Africa (Table 3.7). In addition, adherence to international environmental standards on carbon dioxide emissions will also place increasing pressure on Middle East oil producers and exacerbate latent tensions between exporters and importers. Conversely, it will likely enhance the international gas trade to the benefit of Iran, UAE and to a lesser extent, Australia (Table 3.7).

The Energy Community Concept

It has been suggested that the global energy market has increasingly moved towards cooperation between producers and consumers over the past thirty years (Bahgat, 2001, 542). Of course, energy cooperation can take various forms and incorporate various non-spatial and spatial elements and can occur at different scales – local, national, international. From a geopolitical perspective, the energy community concept holds considerable long-term promise for the maximisation of energy security among regionally cooperating states.

An energy community can be defined as an energy network which is designed to enhance energy cooperation and which is aimed at maximising long-term certainty in energy demand and supply among contiguous and other regionally cooperating states. It is likely to comprise a set of specific goals relevant to the individual and collective needs of participating states which will lead to the construction of a set of mutual obligations and rules which legally bind states to collective actions (Nag, 2004). In one sense, it can also be seen as another element which contributes to the development or maintenance of economic regionalism.

Already, there exist a few examples of actual or proposed regional energy communities. As noted earlier, in essence, the United States hemispheric policy represents a nascent energy community as is the Asia-Pacific gas pipeline project stretching from Australia to Russia (Miyakawa, 1996, 73). In December 2004, the 25 EU member states and 11 countries of Southeast Europe agreed in Athens to establish an energy community for the whole of Europe. Among other things, this will create a single regulatory authority for the production and supply of energy and it is expected that a new European Energy Treaty will be signed before the end of 2005 to codify the details of this novel arrangement. An energy community for Asia has also recently been proposed which is designed principally to ensure security and sustainability of energy supply (Nag, 2004). Indian Ocean states should take due note of these developments and use them as a basis to enhance regional energy security.

Conclusion

Competition for control over energy resources in the Indian Ocean Region is likely to become increasingly intense and lead to greater inter-state conflict. Concern over climate change will mean that increasing pressure will be placed upon regional states with high carbon dioxide emissions (absolute and per capita) to change their patterns of energy consumption, especially those which are coal and oil dominant (ICCT, 2005). At present, nuclear power is a negligible regional source of electricity although energy dependence and vulnerability may increase its attractiveness in the future. Regional resource holders will assume an even greater sense of geopolitical importance with increasing competition especially among the Indian Ocean's energy-niche economies. In the long run, in order to maximise energy security and environmental security and to minimise the potential for increased inter-state conflict, regional states not only need to pursue policies of energy diversification particularly towards renewable sources, but must also seriously consider collective policies which are aimed at the creation of an Indian Ocean Energy Community (IOEC).

Acknowledgement

Much of the background research for this Chapter was undertaken during a Visiting Research Professorship at Kyoto University in the latter part of 2003. I am extremely grateful to the University and especially to Professor Akihiro Kinda and to all of his colleagues in the Department of Geography for making this visit stimulating, enjoyable and productive.

References

Bahgat, G. (2001), 'United States energy security', *The Journal of Social, Political and Economic Studies*, Vol. 26(3), pp. 515-542.

Balmaceda, M. M. (1998), 'Gas, oil and the linkages between domestic and foreign policies: the case of Ukraine', *Europe-Asia Studies*, Vol. 50 (2), pp. 257-286.

Barnett, J. (1997), 'Review of the 1996 Trilateral Commission Report', *Australian Journal of International Affairs*, Vol. 51 (3), pp. 442-3.

Barnett, J. (2001), *The Meaning of Environmental Security* (London: Zed Books).

BPSRWE (2003), *BP Statistical Review of World Energy 2003*.

Bracken, P. (2000), *Fire in the East: The Rise of Asian Military Power and the Second Nuclear Age* (New York: Perennial).

Brundtland, G. H. (1990), *Our Common Future*, Australian Edition (Melbourne: OUP).

Brzezinski, Z. (1997), *The Grand Chessboard: American Primacy and its Geostrategic Imperatives* (New York: Basic Books).

Calder, K. E. (1996), 'Asia's empty tank', *Foreign Affairs*, Vol. 75 (2), pp. 55-69.

Clingendael International Energy Programme (CIEP) (2004), *Study on Energy Supply Security and Geopolitics* (The Hague: DGTREN).

Cutler, R. M. (1999), 'Cooperative energy security in the Caspian region: a new paradigm for sustainable development', *Global Governance*, Vol. 5 (2), pp. 251-271.

Dupont, A. (2001), *East Asia Imperilled: Transnational Challenges to Security* (Cambridge University Press).

Fesharaki, F. (1999), 'Energy and the Asian security nexus', *Journal of International Affairs*, Vol. 53 (1), pp. 85-99.

International Atomic Energy Agency (IAEA) (2000), Reference Data Series, web site, www.iaea.org/programmes/a2/.

International Climate Change Taskforce (ICCT) (2005), *Meeting the Climate Challenge* (London: The Institute for Public Policy Research).

International Energy Agency (IEA) (2003), *Key World Energy Statistics*.

Johnson, C. (2004), *The Sorrows of Empire: Militarism, Secrecy, and the End of the Republic* (New York: Metropolitan Books).

Kelly, A. (2004), 'ExxonMobil's 2004 energy outlook', available at http://www.exxonmobil.com/corporate

Klare, M. T. (2002), *Resource Wars: The New Landscape of Global Conflict* (New York: Henry Holt).

Martin, W. F., Imai, R. and Steeg, H. (1996), *Maintaining Energy Security in a Global Context* (New York: The Trilateral Commission).

Miyakawa, Y. (1996), 'Mutation of international politico-economic structure and the development of the Pacific maritime corridor in the East Asia orbit', in Rumley, D, Chiba, T., Takagi, A. and

Fukushima, Y., eds., *Global Geopolitical Change and the Asia-Pacific: A Regional Perspective* (Aldershot: Ashgate), pp. 49-75.

Nag, B. (2004), 'Towards an Asian energy community: an exploration', paper given to a Conference on 'Asian economic integration: vision of a new Asia', Tokyo, November.

National Energy Policy Development Group (NEPDG) (2001), *National Energy Policy* (US Government).

US Energy Information Administration.

Wilkinson, R. (2003), 'The oil patch', *Oil and Gas Viewpoint*, 14 May, pp. 17-18.

Williams, R. (2005), 'Uranium mine is inevitable, says stakeholder', *The West Australian*, 7 January, p. 14.

THE SECURITY
SPECTRUM

CHAPTER 4

Energy Transportation in the Indian Ocean: Geostrategic and Security Considerations

Vijay Sakhuja

In an age where countries compete on a global scale, the selection and accomplishment of a trade and transportation security strategy is a complex exercise. It can have a dramatic impact on a state's economy. Depending upon the level of international trade, countries take actions and make investments in security measures that will bear fruit if they are applied in concert with national interests. At the same time, the resurgence and threat of sea piracy and global terrorism around the world has put a special emphasis on trade and transportation security. Growing threats and modes of attack have forced governments and the maritime shipping community to find new ways to respond to these threats.

This chapter therefore attempts to highlight the geostrategic and security requirements of energy transportation strategy. In particular, it examines the energy trade and transportation strategies adopted by China, India, Japan and the United States in the Indian Ocean.

Geostrategic Considerations

From ancient times, states have developed trade and transportation security strategies based on geostrategic considerations. Interestingly, geostrategy is the fulcrum on which transportation strategies and security operations are planned. The economists and politicians highlight geostrategy in their discussions as it helps them to understand and appreciate strategic relationships and requirements. To an economist, the shortest route, low transportation cost and timely delivery of cargo are some of the factors that play a dominant role in developing a maritime transportation strategy. To a politician, it is the state of relations with countries located along the transportation route that will help develop a transportation strategy for the growth of its maritime enterprise. When the naval and maritime forces develop transportation strategies, the location of friends and adversaries as well as the geography that has to be traversed to get to the assistance of the other is taken into consideration. It is beyond doubt that it is the appreciation of geostrategy and its effects on transport that will determine the strategy.

A geostrategic region can be defined as a space that encompasses a state's perceived political, economic and military interests. These interests are important enough for a state to use all possible means available to it to protect its national interests. Geostrategic regions vary in size, numbers, and location and primarily depend upon the interests (global or regional) of a nation-state. In the case of a state with global interests, there may be more than one geostrategic region. The United States, which is a global power, has interests in both the Indian and the Pacific Oceans and considers both of these oceans as geostrategically important. Similarly, the Persian Gulf is important to China, India, Japan and the United States for their energy requirements and therefore these countries have developed close diplomatic relations with several countries in the region. On the other hand, for a state with only local interests, the geostrategic region may only be its territorial séa or the Exclusive Economic Zone (EEZ). Such states exercise their influence in their area of interest. For instance, Fiji, a small island state, considers its EEZ as its area of geostrategic interests and has a smaller navy to safeguard its maritime and national interests. States that need to influence events far beyond their EEZs, establish more extended geostrategic boundaries depending upon their perceived interests. They build naval forces that can traverse long distances to safeguard national interests. Britain, with its interests in Hong

Kong (before 1997), had maintained a permanent naval presence in the area. Straits and choke points can also be considered as part of a geostrategic region. However small in size, they are of equal strategic importance to the state which controls them. For Singapore and Indonesia, the Straits of Malacca are important enough to consider undertaking regular patrolling of the area.

Sea lines (or lanes) of communication (SLOC) are geostrategic regions. In brief, a SLOC is a route taken by a ship to transit from point A to B. In maritime terms, it should be short, economical and safe for transporting cargo. During times of peace, SLOCs serve as commercial trade routes but during war these routes are considered strategic ways. During the Cold War, there was significant apprehension over the security of the SLOCs that funneled into chokepoints due to Soviet military threats. For instance, the Soviet Union had deployed submarines and *Backfire* bombers based in southern Afghanistan to challenge the sea lanes in the eastern Mediterranean, Red Sea, and Indian Ocean. Similarly, Moscow wanted to obtain military facilities in Mozambique to enable it to oversee the maritime traffic transiting through the Mozambique Channel, a critical chokepoint hosting the mineral trade from East Africa to Europe–United States. In 1968, the Soviet Union had established an Indian Ocean squadron comprising fifteen to twenty-five ships that had access to naval facilities in Vietnam, Syria, Libya, Ethiopia, the People's Democratic Republic of Yemen (South Yemen), and Seychelles. It also began to fly ocean reconnaissance flights and to maintain forward deployments. These initiatives represented the Soviet resolve to challenge US SLOCs. Besides, these deployments were symbols of Soviet naval power and its interests in the Indian Ocean. Similarly, despite the end of the Cold War and the disintegration of the erstwhile Soviet Union, Moscow had three overseas bases — Camh Ranh (Vietnam) and Tartus (Syria) naval bases together with the Lourdes listening post in Cuba. The Camh Ranh lease was signed in 1979 and expired in 2004. Even in the post Cold War period, sea lane and chokepoint security continues to be as critical as it was during the Cold War. The threat has gained greater significance because of the globalization of the world economy and the greater dependence on foreign trade for economic vitality.

Security Considerations

In this age of terror, it is widely accepted that maritime infrastructure has emerged as the soft underbelly of states that can be attacked with little effort. The threat of terrorism looms large in the maritime domain and has the capability to disrupt the free flow of commerce. Tanker shipping appears to be the most vulnerable and has the potential to be the achilles heel of sea borne trade. Tankers can be used as a means of delivering weapons of mass destruction and have the capacity to disrupt and even destroy maritime enterprise and threaten peaceful use of the seas. Notwithstanding these threats, the tanker transportation system has emerged as the most convenient and cost effective mode of transporting large volumes of energy resources, be it crude oil, LNG, CNG and other petroleum products. Nearly 90% of the world's energy cargo movement takes place via tankers and with over 60% of the world's oil shipped onboard more than three thousand tankers. One of the biggest challenges facing security agencies is how to safeguard the energy sea lane. It is generally agreed that terrorists have the capacity and capability to deliberately crash a ship or a boat into a supertanker in any chokepoint. Such an act will result in explosion and spread the burning oil in the channel that can shut down the channel for a long time. This is bound to impact heavily on global markets, sea trade and the maritime insurance industry. Being large, the tankers have limited maneuverability and are therefore the chosen targets. For instance, Al Qaeda demonstrated this limitation by attacking the tanker M V Limburg. A statement following the attack warned that it "was not an incidental strike at a passing tanker but . . . on the international oil-carrying line in the full sense of the word" (Luft and Korin). Reportedly, in June 2002, the Moroccan government arrested a group of Al Qaeda operatives suspected of plotting raids on British and American tankers passing through the Strait of Gibraltar (Luft and Corin).

The threats to energy sea lanes also arise from piracy. In recent times, the distinction between piracy and terrorism seems to be fast eroding. For instance, in August 2001, the M V Ocean Silver, while transiting through the Malacca Strait, was seized by Aceh rebels. The six crewmembers of the vessel were taken hostage. The rebels issued a warning that all ships transiting through the straits between Sumatra Island and Malaysia must first get permission from the insurgents. Earlier, in May 2001, Thai police captured a consignment of 15,000 bullets, grenades, landmines and TNT explosive devices destined for the Aceh rebels (Mahmood).

Similarly, in another incident, the merchant vessel M V Cordiality, was seized by the LTTE near the port of Trincomalee and 5 Chinese crew were killed. In August 1998, a Belize-flagged general cargo vessel, Princess Kash, was hijacked by the Tamil Tigers. While on its way to Mullaitivu, a Tamil stronghold, the Sri Lankan Air Force bombed the vessel to prevent the ship's cargo falling into the hands of the rebels. The status of the 22 crew members is still not known.

However, strategists differ in their understanding of terrorism and piracy. According to Brian Jenkins, an expert on terrorism and security, it is incorrect to conclude that an increase in piracy will result in an increase in the terrorist threat. He also notes that there is no indication that terrorists and pirates are operating in close cooperation. Their aims are different (Gatsiounis, 2004). Similarly, Captain Mat Taib Yasin, a former Royal Malaysian Navy Officer, and Senior Fellow at the Maritime Institute of Malaysia (MIMA), noted that "The pirate wants to enjoy his loot. The terrorist wants to destroy the enemy, get political mileage – and he's prepared to die" (Gatsiounis, 2004). It has also been argued that "How far little money might go in convincing pirates to assist terrorists. After all, money – loot – is – the pirate's language, and the terrorists who may be eying the Strait are well funded" (Gatsiounis, 2004). Similarly, Adam Young and Mark J. Valencia make clear the distinction between piracy and terrorism by noting that:

> Terrorism is distinct from piracy in a very straightforward manner. While piracy is a crime motivated by greed and thus predicated on immediate financial gain, terrorism, and its maritime manifestation, 'political piracy', or maritime terrorism, is motivated by political goals beyond the immediate act of attacking or hijacking a maritime target. Terrorism at sea includes the twin threats of attacks on shipping and the threat of ships being used as weapons, and the threat of ships being used to deliver concealed weapons

of mass destruction (in containers or within the ship's superstructure). Both have the potential to cause systemic economic dislocation. Indeed, the effect of a major attack on a US port or on a transshipment hub such as Singapore would be felt globally. For example, the aspect of tanker transportation most economically sensitive to terrorism is insurance. Oil and liquefied natural gas (LNG) supplies may be adversely affected by a sharp increase in insurance premiums, or the withdrawal of coverage altogether from areas judged to be high risk. (Young and Valencia, 2003).

Energy Demands of China, Japan, India and United States

China

China became a net importer of oil in 1994. By 2000, the gap between domestic production and national consumption touched 1.591 million barrels of oil (mbo) per day. By 2049, Chinese crude oil demand is estimated to touch about 10 mbo per day unless additional proven reserves are discovered domestically. Though coal is in abundance, it is of poor quality, and hydro power is not sufficient to meet the growing demands of economic modernisation. Energy shortages are a bottleneck for China. It will have to import both crude oil and gas to sustain its economy.

China imports sixty percent of its oil from the Persian Gulf region and therefore energy security is uppermost in the Chinese government's mind. There is a worry among officials that domestic supply can only last a few days and oil reserves are crucial to buffer the economy against import disruptions. Currently, there are very few oil reserves and the national government has outlined a plan to build reserves in a bid to create energy security. According to one proposal, China will procure oil from foreign sources and store it in tanks in the southeastern coastal Provinces. The stockpile could be of the order of 6 million tons by 2005.

China imports a large proportion of its oil supplies from the Middle East and Southeast Asian countries. Among these, the bulk of the oil is obtained from Iran, Iraq and Libya. The Straits of *Malacca, Hormuz* and *Suez* are strategically important to Chinese sea lane security.

Table 4.1. China's Energy Security Context.

Proven Oil Reserves (1/1/04E):	18.3 billion barrels
Oil Production (2003E):	3.54 million barrels per day (bbl/d)
Oil Consumption (2003E):	5.56 million bbl/d
Net Oil Imports (2003E):	2.02 million bbl/d
Crude Oil Refining Capacity 1/1/04E):	4.5 million bbl/d
Natural Gas Reserves (1/1/04E):	53.3 trillion cubic feet (Tcf)
Natural Gas Production (2002E):	1.15 Tcf
Natural Gas Consumption (2002E):	1.15 Tcf

Beijing has long been aware of the geostrategic importance of Myanmar. The strategic infrastructure built by the Chinese includes roads, communication and intelligence networks, and military bases. Another strategic development in Myanmar is the mushrooming of China's electronic intelligence systems. Among these, the most important is the maritime reconnaissance and electronic intelligence system on the Great Coco Island in the Bay of Bengal. This reflects Beijing's strategic interest in the Bay of Bengal. Further, the intelligence network in Sittwe and Zedetkyi Kyun off the Terrasserim coast in South Myanmar provides China with its desired strategic surge into the Indian Ocean. The facility at Zedetkyi Kyun is especially sensitive and the Chinese now sit astride the Malacca Straits to monitor any maritime contingency affecting its sea lane security.

Similar Chinese initiatives can also be seen in Pakistan. In March 2002, the visiting Chinese Vice-Premier, Wu Bangguo, laid the foundation stone of a deep-sea port at Gwadar. The first phase of the project would be completed in three years at a cost of US $248 million, of which US $198 million would be contributed by China. This involves construction of three berths of 200 meters each, with 300 meters of back-up area. On completion of the second phase, the port would be able to accept dry cargo vessels of 100,000 dead-weight tonnage and oil tankers of 200,000 dead weight tonnage. Gwadar would eventually provide facilities for sea borne trade and the transshipment of cargo, including gas from/to the Central Asian states and Xinjiang in China.

China is assisting Pakistan to develop Gwadar with the strategic aim of watching its sea-lane from the Persian Gulf. In pure geographical terms, Gwadar is of strategic importance for China and helps it to sit astride the sea lane originating from the strategic choke point of Hormuz. The Chinese have denied that Gwadar has any military dimensions for China. According to the Chinese foreign ministry spokesman, Sun Yuxi, Gwadar is a civil port and there is no (Chinese) naval involvement. As a matter of fact, Beijing's interests in Gwadar go well beyond the safety of its energy-related shipping from the Persian Gulf and provides it with a forward base to monitor US naval activity in the Persian Gulf region as well as Indian naval activity in the Arabian Sea. A quick analysis of these developments clearly indicates that China has followed an aggressive strategy for sea lane security, which has been flavoured with economic, political and military components.

Beijing's military sales to Tehran, especially in the field of weapons of mass destruction (WMD) and related technologies, have enabled it to establish a foothold in the Iranian port of Bandar Abbas, well inside the Persian Gulf and astride the Strait of *Hormuz*. In 1987, China sold to Iran "Silkworm" anti-ship cruise missiles and also agreed to provide Iran with equipment and know-how to develop and test medium-range ballistic missiles. By 1990, China and Iran reportedly signed a ten-year agreement for scientific cooperation and the transfer of military technology. In 1996, Iran test-fired a Chinese origin C-802 surface-to-surface cruise missile from one of its 10 Chinese-built Houdong class patrol boats. China has thus emerged as the main source for equipment and know-how to develop and test Iranian medium-range ballistic missiles.

A relatively new entrant in the Chinese security calculus is Maldives. Since the establishment of diplomatic relations in 1972, economic and trade cooperation between China and Maldives has been developing continuously. The scale of Sino-Maldivian economic cooperation is quite limited with a bilateral trade volume of US$1.358 million. However, what is of greater interest is that China has "engineered a manner of a coup by coaxing Maldives' Abdul Gayoom government to let it establish a base in Marao". According to media reports, the deal was finalised during the Chinese Prime Minister Zhu Rongji's visit to Male in May

2001. In geostrategic terms, this development is indeed noteworthy and would enable China to monitor its sea lanes that run along the Indian coast. Besides, it would be able to monitor US naval activity at Diego Garcia and also engage in electronic surveillance in the area, a development similar to its electronic surveillance initiatives in Kiribati in the Pacific Ocean.

As noted, China has long understood the strategic importance of the Indian Ocean. The Indian Ocean figured in the strategic thinking of Chinese ancient mariners who sailed in these waters for trade. Chinese strategists, maritime planners and practitioners are convinced that the Indian Ocean dominates the commercial and economic lifelines of the Asia-Pacific region, and this reality is of increasing importance to China. China's naval and military surge into the Indian Ocean is a major strategic priority for planners in Beijing. The surge is aimed at consolidating the Chinese energy sea lane as well as a military posture in an emerging hostile environment wherein a possible clash with the US in the foreseeable future cannot be ruled out.

One of the primary missions of the PLA Navy is safeguarding sea-lanes. China relies heavily on sea-borne trade for its economic vitality as well as for its energy needs. In the event of the closure of the strategic choke points of the Persian Gulf, South East Asia and the South China Sea Strait, China will be adversely affected. The PLA Navy has been tasked to keep the sea lanes open and escort Chinese tankers transiting the strategic straits.

Towards that end, China has developed an aggressive strategy aimed at assisting Myanmar, Maldives, Pakistan and Iran to build naval facilities. Importantly, much of the assistance is concentrated along the sea lane from *Malacca* to *Hormuz*. The most alarming aspect of these developments is the likely use of these facilities in support of the PLA Navy's surge into the Indian Ocean. The present Chinese commitment in infrastructure build-up, impressive as it is, is probably the beginning of a long-term Chinese interest in the Indian Ocean.

It has been argued that, "China is building strategic relationships along the sea lanes from the Middle East to the South China Sea in ways that suggest defensive and offensive positioning to protect China's energy interests, but also to serve broad security objectives" (US Director of Net Assessment, 2005). The report noted that that the "string of pearls" strategy includes a new naval base under construction at the Pakistani port of Gwadar, naval bases in Myanmar, a military agreement with Cambodia, strengthening ties with Bangladesh and an ambitious plan under consideration to build a 20-billion-dollar canal in Thailand to bypass the Strait of Malacca. The report also noted that China, by militarily controlling oil shipping sea lanes, could threaten ships, "thereby creating a climate of uncertainty about the safety of all ships on the high seas …China is looking not only to build a blue-water navy to control the sea lanes, but also to develop undersea mines and missile capabilities to deter the potential disruption of its energy supplies from potential threats, including the US Navy, especially in the case of a conflict with Taiwan" (US Director of Net Assessment, 2005).

Japan

During the last two decades, Japan has taken a series of steps to safeguard its energy security. These include building a national flag fleet of ships capable of transporting oil and gas from the Persian Gulf, undertaking maritime exercises with maritime forces of countries along its energy sea lane and deploying its maritime forces as far as the Arabian Sea.

Table 4.2 Japan's Energy Security Context

Proven Oil Reserves (1/1/04E):	59 million barrels
Oil Production (2003E):	120,000 barrels per day (bbl/d), of which 5,000 bbl/d was crude oil
Oil Consumption (2003E):	5.57 million bbl/d
Net Oil Imports (2003E):	5.45 million bbl/d
Crude Oil Refining Capacity (1/1/04E):	4.7 million bbl/d
Natural Gas Reserves (1/1/04E):	1.4 trillion cubic feet (Tcf)
Natural Gas Production (2002E):	0.10 Tcf
Natural Gas Consumption (2002E):	2.67 Tcf
Net Natural Gas Imports (2002E):	2.57 Tcf

In 1981, Hideo Sekino, a much respected defence commentator, argued that the protection of sea lanes be given priority in the national defence of Japan. He noted that prevention of direct invasion be reordered as a secondary function of the Japanese navy. Securing the sea lanes north of Indonesia was considered to be strategically important. According to Sekino, a *guerre de course* was the most likely form of conflict. In 1977, the Director General of the Japanese Defence Agency publicly stated that Japan should defend "key transport lanes within 1000 miles of Japanese coast".

In May 1981, following a summit meeting with President Reagan, Prime Minister Zenko Suzuki declared that Japan had agreed to take responsibility for defending sea-lanes up to 1000 nautical miles. In 1983, the Japanese Prime Minister, Yashuhiro Nakasone, while on a visit to the United States, said that Japan should become "an unsinkable aircraft carrier" and also be able to control the Sea of Japan straits. He articulated the notion of sea lanes as "between Guam and Tokyo and between the Strait of Taiwan and Osaka." The 1983 White Paper on defence included an explanation of this policy and its requirements.

In recent years, Japan's military has increasingly begun to bridge the gap between being a self-defence force and a regional military power. The transformation was formalized in 1997 when the US-Japan Defence Guidelines were revised and the annual Defense White Paper publicly addressed a regional strategy in "areas surrounding Japan". In 1999, while taking part in US exercises in Guam, Japan deployed fighter aircraft outside its territory for the first time since World War II. In 2000, the JSDF carried out exercises with other regional forces without the United States.

In February 2000, Japan's Foreign Ministry announced that Tokyo was considering deploying vessels to patrol the Straits of Malacca where shipping has been plagued by piracy. Maritime Safety Agency (coast guard) ships were expected to form part of the multinational anti-piracy patrol comprising coastguard and naval vessels of China, Republic of Korea, Malaysia, Singapore and Indonesia. Strong protests and non-acceptance of the proposal by China resulted in Japan keeping the idea on hold. In 2001, Japan, for the first time sent, officers to observe Cobra Gold, a joint Thai-US military exercise, the largest in the region.

By participating in the US-led war on terrorism in Afghanistan, Japan has redefined its maritime defensive perimeter out to three thousand nautical miles from the homeland. Although the 1000 miles limitation has continued to exist to date and is noted in white papers, the counter terrorism proposal has only supported the growing debate in Japan on revising the Constitution. Distant and joint naval operations once unthinkable are now accepted as part of national defence. If, at any stage, the Japanese agree to assist the US in its war against the 'axis of evil', the defensive perimeter will extend far beyond 4000 nautical miles into the Persian Gulf.

The changing role clearly reflects the creeping assertiveness of the Japanese navy and a desire to shed associated symbolic pacifism. A sophisticated strategy has been developed by the Japanese to incorporate new roles into the military strategy. It has all the elements of patience and detailed planning. It begins by developing new missions, training, and then perfecting it and presenting it for public acceptance. In the past, Japanese maritime forces have been engaged in several incremental military activities (overseas deployment in 1958, sea-lane protection out to 1000 nautical miles in 1983, mine sweeping in Gulf War 1991, UN Peace Keeping in Cambodia and Mozambique between 1992 and 1995) which fall far beyond the limits of the Japanese pacific constitution. These have been undertaken without inviting strong international and domestic opposition.

However, what is more important is the Japanese decision to dispatch naval forces to the Indian Ocean in support of the US-led war on terrorism. This decision has only added to regional concerns of Tokyo's roles both in terms of expanded military reach and the type of missions. The vessels deployed in the Arabian Sea (two destroyers and a support vessel that returned after five month deployment) can in no way be termed defensive. These vessels are inherently power projection platforms. More importantly the deployment has augmented the Japanese ability to undertake energy sea lane defence as far as the Strait of Hormuz.

Since the mid-1990s, India and Japan have been discussing common security concerns that relate to nuclear policies and maritime security issues (Mathur, 2002). Japan is heavily dependent on Gulf oil for its energy requirements and its vessels have to transit the Andaman Sea-Malacca Straits and the Arabian Sea before setting course either to enter or exit the Strait of Malacca (Noer, 1996, 18). With regard to maritime cooperation, both Japanese and Indian war ships have visited each other's ports and participated in the International Fleet Reviews. In 1995 and in 2000, Indian warships visited Tokyo and Sasebo, respectively. This was in response to an earlier visit of a Japanese naval ship in 1995. Another area of mutual interest between the two maritime forces is in search and rescue at sea.

Both India and Japan have taken a strategic initiative to assert greater influence in the Asia Pacific region. As noted, India held naval exercises in the South China Sea in 2000, while Japan has stretched its area of defence out to 2,000 miles towards the Malacca Strait. Because of their geographic locations, the two countries do not appear to compete with each other for influence. Instead, their strategic geographic areas overlap, covering a large proportion of the South China Sea. This clashes with Chinese interests in the same area, and from the Indian and Japanese perspectives they should contain Chinese influence in the region.

Japan is expanding its influence from Northeast Asia to Southeast Asia by gaining access to Singapore's naval facilities for staging operations in the Malacca Strait to challenge any Chinese initiatives in the region (Weekly Global Intelligence Update, 2000). Tokyo and Singapore concluded a bilateral military agreement in 2000 that permits Japanese patrol ships and aircraft to be stationed in Singapore to evacuate Japanese nationals in disturbed areas as well as to assist UN peacekeeping operations in Southeast Asia. Earlier, during the Indonesian crisis leading to the fall of President Suharto, Tokyo had positioned aircraft and patrol vessels in Singapore on standby for evacuation duties. This agreement increases Japan's naval presence in the Malacca Strait and South China Sea, as far as two thousand miles from home.

According to J. Mohan Malik, the Director of the Defence Studies Programme at Deakin University in Australia, the main aim of China's Asia policy is to prevent the rise of a potential rival or a competitor. Towards that end, Beijing's strategy is aimed at "restraining Japan and containing India" (Malik). He notes that "what distinguishes China from its near-rivals, Japan and India, is its permanent membership of the UN Security Council and declared nuclear weapons state (NWS) status, making it a far more important player in international forums and the sole Asian negotiating partner of the United States on global and regional security deliberations." In order to counter the influence of India, Beijing has developed close economic and military relations with Pakistan, Sri Lanka, Bangladesh and Myanmar and aims to negotiate from a position of strength. As regards Japan, the Chinese strategy is aimed at propping up North Korea to challenge any Japanese regional ambitions (Malik).

India

India's annual consumption of petroleum products for 2002-3 was 111.5 mmt, with gas accounting for another 30 mmt oil equivalent. About 67 percent of India's energy need is sourced from the Persian Gulf and 17 percent from West Africa. There is very little that is sourced from Southeast Asian countries. However, keeping in mind the vulnerability of energy security both in terms of sourcing and transportation, India may, in the future look towards Southeast Asian markets. Such supplies would have to transit through the Straits of Malacca. Patrolling in the Malacca Straits is not new to the Indian Navy. In April 2002, as part of their bilateral cooperation, Indian and US naval ships engaged in joint escort duties in Malacca Straits waters that are home to both pirates and terrorists. The Indian naval ship, *Sharda*, had assumed responsibility from the *USS Cowpens* to escort American commercial vessels carrying 'high value' goods transiting through the strait. According to a bilateral arrangement, US naval vessels patrolled sea areas in Southeast Asia while the Indian Navy concentrated in the Bay of Bengal and the Indian Ocean. This initiative emerged as an outcome of the understanding between New Delhi and Washington to revive the Malabar series of joint naval exercises suspended as fallout from the 1998 Indian nuclear tests. It was also agreed to recommence joint operations that include search and rescue exercises to help vessels in distress in the Indian Ocean, Bay of Bengal and Arabian Sea, safety of sea-lanes, anti-piracy and issues relating to maritime order at sea.

The Indian Navy has bilateral maritime agreements with several navies in its neighbourhood. For instance, it has the 'IndoIndon' bilateral agreement with Indonesia that involves patrolling the western approaches of the Malacca Straits. In

September 2004, the Indian navy conducted a cooperative exercise called 'IndIndoCorpat' (Indo Indonesia Coordinated Patrol) in the region with the Indonesian Navy.

Similarly, the Indian and Singapore navies have a highly developed maritime cooperation agreement that includes joint naval exercises, submarine training and bilateral exchanges. India and Singapore have been holding joint exercises for the past 11 years and these have been in Indian waters. Rear Admiral Ronnie Tay visited New Delhi on an official visit in September 2004 and has invited Indian Naval warships to undertake joint manoeuvres in the South China Sea. As part of India's Look East policy and the growing nature of Indo-Singapore naval relations, Tay was expected to take forward the proposal at the Bay of Bengal Initiative for Multi-Sectoral Technical and Economic Cooperation (Bimstec) meeting in Bangkok in August 2004 that involved joint patrolling of busy sea lanes of communication by the Indian Navy with regional countries. India-Singapore Air Force cooperation has now graduated to joint exercises with the Singapore Air Force who will deploy six F-16 aircraft for the exercises to be held in the skies over Kalaikunda, Gwalior and Pokhran.

The Chief of the Royal Malaysian Navy, Admiral Dato Seri Mohammad Anwar Bin HJ Mohammad Nor, during his visit to India met the Indian Navy chief. The two sides discussed the issue of security in the Straits of Malacca and also explored opportunities for the training of Malaysian naval personnel and navy-to-navy cooperation on the high seas in the Indian Ocean Region. India and Thailand are also negotiating similar agreements.

United States of America

According to the US 2001 National Energy Policy (NEP), domestic oilfield production will decline from about 8.5 million barrels per day (mbd) in 2002 to 7 mbd in 2020, while consumption will jump from 19.5 mbd to 25.5 mbd. The shortfall will be through imports or other sources of petroleum, such as natural gas liquids, and will have to rise from 11 mbd to 18.5 mbd (Klare, 2002). The then Secretary of Energy, Spencer Abraham, told a National Energy Summit on March 19, 2001 that "America faces a major energy supply crisis over the next two decades . . . The failure to meet this challenge will threaten our nation's economic prosperity, compromise our national security, and literally alter the way we lead our lives" (Klare, 2002).

The United States currently obtains only about twenty percent of its imported petroleum from the Persian Gulf. Until recently, higher level of production have enabled world oil prices to remain relatively low thus benefiting the US economy. According to the NEP, with domestic production in decline, the Persian Gulf "will remain vital to US interests". The US has been actively engaged in the politics of the Persian Gulf, be it the 1991 Iraq war or the 2003 war against Saddam Hussein.

During World War II, president Franklin D Roosevelt signed an agreement with Abdul-Aziz ibn Saud, the founder of the modern Saudi dynasty, to protect the royal family against its internal and external enemies in return for privileged access to Saudi oil (Klare, 2002). Similar assurances were also given to the Shah of Iran (the link was severed in 1980) and to the leaders of Kuwait, Bahrain and the United Arab Emirates. In pursuance of its energy security policy, the US has followed an aggressive strategy to supply military hardware to the Persian Gulf countries. There also exist formal agreements for military exercises, training and even policymaking. US naval vessels are a common sight in the Persian Gulf with the CENTCOM headquarters in Bahrain. The US has followed intrusive policies, even shaping the domestic politics of these friendly states. Consequently, it has exposed itself to increased risk of involvement in local and regional conflicts. This strategy has had an adverse impact on its relations with the major oil-producing states who now perceive the US as a 'scrounger' of their oil assets. The US used force in 1987 and 1988 to protect Kuwaiti oil tankers from Iranian missile and gunboat attacks, and then in 1990 and 1991 to drive Iraqi forces out of Kuwait.

In January 2005, there were reports that the U.S. Navy diverted at least 12 unarmed ships of the Navy's Military Sealift Command and some supply ships carrying critical war materiel away from the Suez Canal (Wood, 2005). The vessels transited around the Cape of Good Hope en route to the Persian Gulf. This was in response to intelligence reports of threats to shipping due to recent rise of Saud Hamud al-Utaibi in al-Qaida's leadership, a maritime terror expert believed to have been responsible for the attacks on the USS Cole and the supertanker MV Limburg. "Al-Utaibi is the new head of al-Qaida on the Arabian Peninsula, and that heightens the threat to shipping certainly within that region" (Wood, 2005).

In April 2004, Admiral Thomas Fargo, Commander-in-Chief, US Pacific Command, (CINCPAC), announced that the US military was planning to deploy Marines and Special Forces troops on high-speed boats in the Malacca Straits to combat terrorism, proliferation, piracy, gun running, narcotics smuggling and human trafficking in the area. The deployment was being conceptualised under the Regional Maritime Security Initiative (RMSI). In response, Malaysia reacted to this initiative and noted that the US should obtain permission from regional countries as it impinged on their national sovereignty. Likewise, Indonesia was also averse to the US initiative and wanted the US to consult regional countries before any effort to fight terrorism in Southeast Asia. Washington is now pushing ahead the Regional Maritime Security Initiative through 'informal contacts with friends and allies in the region'

Concluding Remarks

Enhanced economic dynamism in the Asia Pacific and the Indian Ocean region, primarily based on energy transportation activity, has resulted in increased sensitivity of the energy sea lines of communications, especially though the Straits of Malacca and the Strait of Hormuz. At this stage, it is difficult to see any 'state-induced threats' to the stability and security of the regions, but states seeking naval facilities in the Indian Ocean are indeed a matter of concern. The geostrategic and economic realities are complex for the entire world community. Some states, particularly the United States, Japan and China, seek an enhanced role in the region for the safety of their long and often vulnerable sea lines of communications.

The economic boom in the Asia Pacific region is dependent on the Indian Ocean. The bulk of the energy requirements for the economic boom must be fulfilled from Indian Ocean states. Presently, there are no indications to suggest that passage of ships through the sea lanes in the region is likely to be disrupted, and the states of the Asia Pacific region will be anxious not to let such an eventuality to take place. Nevertheless, circumstances change and newer challenges may emerge. The obstruction or closure of these lanes will have a major economic and security impact for China, Japan and United States and for the region as a whole. In the event of such a situation, maritime traffic will have to sail further, placing increasing demand on vessel capacity and increasing the time and cost of transportation.

It is also evident that energy security is being threatened and challenged from several directions. The Indian Ocean region has become a safe haven for gunrunning, drug trafficking, illegal fishing/poaching, piracy, terrorism and other maritime crimes and acts that have the potential to disrupt maritime enterprise. These activities have the potential to escalate into skirmishes among regional navies when engaged in safeguarding their maritime interests as well as when operating to safeguard their respective energy sea lanes.

References

Gatsiounis, J. (2004), "Malacca Strait: Target for terror", *Asia Times*, August 11, 2004 available at http://www.atimes.com/atimes/Southeast_Asia/FH11Ae02.html

Klare, M. (2002) , ' US: Procuring the World's Oil', available at http://www.atimes.com/atimes/Global_Economy/FD27Dj02.html

Luft, G. and Anne Korin, "Terror Next Target", *The Journal of International Security Affairs* available at < http://www.iags.org/n1216041.htm>

Mahmood, K., Thailand cracks down on arms smuggling to Aceh http://islamonline.net/English/News/2001-05/18/article4.html

Malik, M. "China's Asia Policy: Implications for Japan and India" at the website of *Virtual Forum on Asian Security* at http://www.arts.monash.edu.au/mai/savirtualforum/PaperMalik 1.htm

Mathur, A. (2002), "Prospect for Indo-Japan Security and Political Cooperation: An Indian Perspective", Paper presented at the JIIA-IDSA Bilateral Dialogue, Tokyo, May 14-15.

Noer, J. H. (1996), *Choke Points: Maritime Economic Concerns in South East Asia* (Washington: NDU Press). In 1993, half of the crude oil tankers entering the Malacca Strait were bound for Japan.

US Director of Net Assessment (2005), "String of Pearls" Military Plan to Protect China's Oil US Report", Washington (AFP) January 18, 2005 available at http://www.spacewar.com/2005/050118111727.edxbwxn8.html.

Weekly Global Intelligence Update, at homepage of *Stratfor* at http://www.stratfor.com dated May 3.

Wood, D. "Terrorism Fears Divert Navy Supply Ships From Suez Canal", Newhouse News Service, January 13, 2005.

Young, A. and Mark J. Valencia (2003), Conflation of piracy and terrorism in Southeast Asia: rectitude and utility. *Contemporary Southeast Asia*, v. 25, no. 2, pp. 269-283.

CHAPTER 5

Non-Traditional Security Issues
Recent Trends in the Hydrocarbon Sector
in the Indian Ocean Region

Aparajita Biswas

The late 20th century has seen a profound change in the concept of security. Non-traditional security issues, especially those related to non-military sources of threat, have assumed prominence. Unlike the past, the nature of the threats and conflicts is undergoing a transformation. The changes associated with the end of the Cold War have been so dramatic and intense that it is reasonable to question whether traditional assumptions regarding the nature of global conflicts will continue to prove reliable in the new post-Cold War era.

The planned aggression against Afghanistan and Iraq by the world's major powers – the United States and UK – was essentially initiated to gain control over valuable ·mineral resources and the enormous oil and gas reserves in the Caspian Basin and the Persian Gulf region. Underlying this aggression is the fact that the USA was losing its 70-year old hegemony over oil in the Gulf region. This was mainly because of two developments. The first was the nationalization of oil production by its producers – the Organisation of Petroleum Exporting Countries (OPEC) – the second factor was the Islamic revolution of Iran in 1979, which resulted in a change in the USA's Gulf allies. However, the final blow came from Saddam Hussein of Iraq who attempted to invade

Kuwait in 1990, and also invited countries such as China, France and Russia for oil exploration in Iraq.

The energy security issue assumes paramount importance in the oceanic regions of the world, which, besides being richly endowed with natural resources, are also a source of a large proportion of the world's production of hard minerals like manganese and hydrocarbons. The heightened world focus on the mineral resources of oceanic regions is a result of two distinct trends. Firstly, a steadily increasing demand for mineral resources and petroleum-based energy sources, and secondly, the rapid progress of technology to a point where huge new and unclaimed resources are now at the disposal of modern industry (Leipzinger and Mudge, 1976, xvi).

The new mineral resources lie below the ocean, extending seaward from the continental shores, where hydrocarbon deposits – oil and natural gas – are trapped in the subterranean strata, to the mid basins where manganese nodules lie in vast collections on the surface of the ocean floor likes rocks on the sea bed (Leipzinger and Mudge, 1976) The oceans contain a number of minerals, but hydrocarbon and manganese nodules are, by far, the two most economically significant categories of minerals.

During the last several decades, the hydrocarbon sector has assumed importance as an area of cooperation among nations, mainly because oil and gas are limited and unequally distributed resources. There is a huge demand for existing reserves of oil by nations aspiring to secure steady supplies. This chapter tries to hypothesise that, while these demands may stimulate valuable cooperation among selected states in the Indian Ocean Region, the seeds of potential tension and conflict are also being sown. Herein lies the contradiction: energy, particularly hydrocarbon, is an area where the dual practice of cooperation and conflict can be discerned. This chapter tries to capture the present trends of this in the context of the Indian Ocean Region.

Potential for Cooperation and Conflict in the Hydrocarbon Sector

At the outset, it is appropriate to recall that, in the autumn of 1973, world affairs underwent a dramatic change. It was a time when petroleum-exporting countries realized that their oil resources could be used as an invaluable source of income and as a means to exert vast economic and political influence, if used tactfully.

As a result, countries came to realise the vital importance of oil in national economic and military security. Industrialised states in particular, belatedly realised the vulnerability of their position, since they had to depend on petroleum-exporting countries for supplies of oil. It became imperative for them to evaluate the situation and recognise the need for new alignments and cooperation with other states, for their own national interests.

Industrialised states considered petroleum-exporting (PE) countries like Saudi Arabia, Iran and Kuwait to be the most coveted partners for cooperation. The PE countries also tried to cooperate with them, with the intention of securing assistance for their own economic and technical development, as well as for military security. This provided the basis for economic and military alliances between industrialised states and the PE countries, with the help of multinational oil companies, and further led to cooperation in the fields of trade and investment. As a result, there was a heavy flow of capital to PE countries, leading to a steady increase in trade and investment. However, there was also a disquieting factor. A considerable portion of the currency inflow into PE countries was used by them for the purchase of weapons. In fact, there are enough data available to demonstrate the close connection between oil and weapons. Over time, this linkage played an important role in unsettling the regional power balance (SIPRI Yearbook, 1974, 62).

Another consequence of the oil-weapons linkage was the development of potential conflict situations in West Asia and the Middle East. The considerable build-up in the arsenal of arms in parts of the Arabian and Persian Gulf region was, therefore, not an isolated phenomenon. It may be noted here that the Indian Ocean Region, which was relatively free from the arms race, had nevertheless become another militarised area during the time of Cold War.

In recent years, the processes of globalisation have redefined the patterns of cooperation and conflict among states. The emerging patterns are showing noteworthy discontinuities from the past, by transforming existing structures. The increasing cooperation in the hydrocarbon sector, because of a shift in consumer patterns from Europe to Asia, is one manifestation of the change. The rationale for cooperation can be ascribed to three broad sets of factors. First, the steadily increasing demand for mineral resources and petroleum-based energy sources, especially in developing states. Second, the amazing technological innovations in sectors like oil and gas pipelines and transportation. These developments led to the setting up of linkages between remote areas richly endowed with hydrocarbon resources, to the mainstream markets. The *energy corridor* concept is a reflection of this reality, as is evident from the 640-km sub-sea pipeline which made it possible for Indonesia to transport its natural gas from South China Sea to Jurong Island in Singapore. Third, with the onset of liberalization and privatization, most of the developing countries have opened up their oil sectors, leading to huge investments from multinational corporations and big local companies.

Patterns of Change

Global Change

Since the 1990s, the world has witnessed a significant shift in world energy consumption patterns. The share of developing countries in global energy consumption has increased considerably, compared to the 1970s. In 1971, Asia – including the OECD Pacific region – accounted for only 14% of the total world demand for energy. Today, its share has doubled, to 28%. In fact, Asia has emerged as the largest oil-consuming region in the world, one percentage point ahead of North America. In 2000, South Asia accounted for approximately 3.9% of the world's commercial energy consumption - up from 2.8% in 1991 (IEA, 2001).

According to a projection of the World Economic Outlook, Asia's share of oil in the global total will continue to increase and touch 35% by 2020. The increase will be evident mainly in China, India and Southeast Asian countries. In volumetric terms, this means that demand for energy in Asia, which was 19 million barrels per day (mbd) in 1997, will grow to over 28 mbd in 2010 and more than 37 mbd in 2020 (IEA, 2001). However, given the limited and declining production of oil in the region, all the incremental demand for oil will have to be met with imports. For example, Southeast Asia is already a net importer of oil, in spite of the presence of oil-exporting countries in the region, like Indonesia, Malaysia and Vietnam. In the case of South Asia, the region consumed around 2.5 mbd of oil, but produced only 0.80 mbd in 2000, making it a net importer of around 1.7 mbd. India and Pakistan account for most of South Asian oil production (i.e. around 7,40,000 mbd). The Middle East is expected to remain the major source of supply of oil to both South and Southeast Asian countries. As a result, oil imports of both these regions are likely to grow sharply, since production remains flat while demand soars (IEA, 2001).

Change in the Indian Ocean Region

The Indian Ocean Region plays an important role in the emerging trends, mainly because the two sea lanes used to transport oil pass through its waters. The first route is around the Cape of Good Hope, onward to Western Europe and the United States, and the other route is eastwards, through the Malacca Strait to Japan. The risks involved in the sea borne transport of oil, especially pipeline transportation, are undoubtedly a deep security concern for many states, involving serious conflict potential. This area is also important because the shipping route from the Middle East to Southeast Asian countries and beyond is a very long one, with most of the oil tankers passing through the Strait of Malacca which is only 3 km wide at its narrowest point. This is the primary sea route from the Middle East and Africa to Japan, Korea, China, Taiwan and other Pacific Rim countries. Any closure of the Strait of Malacca would mean disruption of freight movement and an increase in sailing time and freight rates. Moreover, piracy and geopolitical friction among the countries of the South China Sea region are two major security concerns. However, the major worry is that oil resources are highly concentrated in areas which are subject to political strife. In fact, supplies of oil could be suspended by oil-exporting countries through collective political intervention. Accidents on high seas, or, for that matter, geopolitical incidents, could temporarily halt oil production or close shipping routes. Added to this is the fact that the oil-producing states are displaying renewed confidence in controlling and managing the international oil market. It therefore becomes vital for countries in the Indian Ocean Region to cooperate among themselves, to keep the shipping routes safe and always open. This calls for constant contact and cooperation among them, both at the technical and political levels.

Natural Gas Pipelines: cooperation and conflict

There has also been a significant increase in the demand for natural gas in Asia. The demand covers both LNG (Liquefied Natural Gas) and pipeline gas. The International Energy Agency projects that demand for LNG in Asia will more than double by 2020. What's more, of all forms of energy, demand for natural gas in Asia is expected to grow the fastest. In South Asia, natural gas reserves are around 58.6 trillion cubic feet (tcf), or about 1% of the world's total. This region consumed and produced around 1.99 tcf of natural gas in 2000. Around 48% of this production is accounted for by Pakistan, another 40% by India, and 17% by Bangladesh. In Southeast Asia, there is also considerable growth in the use of pipeline gas, primarily in the member countries of ASEAN, with Indonesia, Malaysia, Brunei, Vietnam and Thailand having significant gas reserves (IEA, 1996). It is important to note that transporting gas is more expensive than transporting oil. This is true for both liquefied gas transported in special carriers as well as gas transported through pipelines, since gas has a low thermal density compared to oil.

In recent years, energy pipeline projects have attracted global capital in the Indian Ocean region. Multinational oil companies such as Unocal, Spie Capag and Total Fina Elf of France, for example, have made significant investments in the Indian Ocean Region. However, while the emerging markets have created opportunities for cooperation in the hydrocarbon sector, they have also created conditions for conflict. There is intense competition among oil firms and nation-states to gain control over precious energy resources like oil and gas, especially in today's era of globalisation. Due to this competition, political processes – both local and global – are continuously shaping the dynamics and complex interrelationship between the oil firms and nation-states on one hand, and NGOs, like environmental and human rights organisations, on the other. At the core of the politics related to the exploration of oil resources are vital issues such as the exploration of new oil fields in virgin areas, smooth management of crude oil flows towards refineries through well-guarded pipelines, enhanced profit revenues through oil extraction, and their likely impact of on environment, development and governance (Rajen, 2002).

The Yadana pipeline project is a case in point and became a focal point of domestic and international debate. This was the first cross-border cooperative pipeline project in Southeast Asia, between the oil companies and the two neighbouring countries of Myanmar and Thailand. The project was jointly owned by Unocal of USA, Total of France, Myanmar's Oil and Gas Enterprise (MOGE), and Thailand's PTT Exploration and Production. Thailand, the principal buyer of the gas, had signed a 30-year contract through PTT, agreeing to purchase Yadana gas from MOGE. This gas field, rated as a world class energy resource, is located 145 meters beneath the Andaman Sea, about 43 miles off the coast of Myanmar. Approximately 220 miles of the pipeline is under water, while the remaining 30 miles traverses the Tenasserim region of Myanmar. The project included development of the Yadana gas field and the construction of a pipeline extending from the offshore field of the Andaman Sea, across Myanmar's remote southern panhandle, to Ban-I-Trong at the Myanmar-Thailand border and onward to Thailand. The Yadana gas field is endowed with reserves of more than 5 trillion cubic feet of natural gas. Production has started and will increase to 525 million cubic feet per day. The gas is used to fuel a 2,800-MW power plant at Ratchaburi, operated by the electric generating authority of Thailand. It is estimated that the pipeline, currently valued at US $1.2 billion, is likely to provide the state an annual revenue of more than US $200 million (www.earthrights).

It is to be noted here that the Yadana Gas pipeline project has been the focal point of debate primarily over the acceptability of foreign investment in Myanmar. The military regime of that country had been internationally condemned both for human rights abuses and for its refusal to cede power to the state's democratically elected government. The project was controversial both as a threat to the environment and for causing the dislocation of local people. The most serious charge against the project was that "slave labour" was being employed and the land of local villagers had been confiscated for the pipeline. Environment-related organisations were apprehensive of the fact that the pipeline which passed through the Tenasserim region, the largest rain forest in the mainland of Southeast Asia, would have an adverse impact on plant and wildlife. Further, it was also home to indigenous groups of people such as the Karen, Mon and Taroyans, who were fighting for autonomy from mainland Myanmar. Human rights activists alleged that, following opposition to the project by the local people, the military government adopted a scorched earth campaign that destroyed entire villages. It also engaged in a pattern of systematic human rights abuses and environmental degradation as it sought to fulfill its contractual responsibilities to Unocal and Total, to provide security for the project. According to the Human Rights report, "Hundreds of thousands of Karens and Mons were taken captive as forced labourers, to build a railway line, roads and new army bases in the vicinity of the pipeline's route". The Myanmar segment of the pipeline was completed in 1997, with thousands of troops guarding the pipeline route. In Thailand too, the Yadana project attracted the wrath of the people, both for environmental and economic reasons. The Yadana project, therefore, symbolises a new trend – collaboration between the global oil companies and neighbouring countries and organised protests against the project at the international level. The campaign against oil investment in Myanmar continued throughout the USA and Europe. The case against Unocal for human rights abuses committed against villagers along the route of the Yadana pipeline, is still underway in the California State court (www.earthrights).

Meanwhile, the "great game" is once again being played in the Caspian Sea basin and Central Asia, this time over access to energy and transit facilities. In 1996, the construction of a US $3.5 billion pipeline from Turkmenistan's Daulatabad gas fields to Pakistan, was proposed by CentGas, a consortium led by Unocal, to construct a 1,635-km-long gas pipeline which would pass through Herat, Kandahar, Quetta and Multan, before entering India, to join the HBJ link of GAIL (Gas Authority of India Limited). It was originally meant to pass through Mazar-e-Sharief, Kabul and Lahore; however, the route was changed to make the extension to India easier. It is interesting to note here that both the USA and Russia are concerned about this pipeline, but for different reasons. Russia's interest is that, without the pipeline, Turkmenistan remains hostage to the Russian gas transportation system as a means to access western markets. This is leverage which Russia is loath to concede. Russia has also figured that, in both the short-term and mid-term, it would be cheaper to obtain gas from Central Asia rather than make huge investments to develop its own gas fields in Siberia. The United States, however, wants to see the pipeline open for Turkmenistan's gas to reach the coast of Karachi, from where it would be shipped to western countries as LNG (Dutta, 2004).

Regional politics also played a major role in defining the game. In mid-2000, the Government of Pakistan affirmed that it would permit a gas pipeline linking Iran's massive gas reserves to India, across its territory. Pakistan would earn transit fees of up to US $1 billion per year for Iranian gas supplied to India, and would also be able to purchase gas from the pipeline itself. While both Iran and Pakistan have shown great interest in this project, India has steadfastly refused to accept any energy lifeline passing through Pakistan so long as political and military tension over Kashmir persists. Realising that the gas pipeline may remain a 'pipedream' (rather than a 'peace pipeline') because of the political turmoil in the region, Iran offered gas (as LNG) in ships at a rate comparable to pipeline supplies. This move of Iran has forced Qatar to lower its rate for gas from the 5.5 million tonnes Dahej project. The point to note here is that Iran desperately needs to modernise its oil sector as US sanctions have blocked all western investment in the country (Dutta, 2004).

There are plans for another overland gas pipeline project linking Bangladesh to India, spearheaded by Unocal, as part of its South Asia Integrated Gas (SAIG) project. Unocal also plans to set up a US $2 billion integrated pipeline connecting Myanmar to the gas fields of Bangladesh, en route to Haldia Port in India. Here, India prefers to wait and watch as Bangladesh's domestic politics remains in turmoil over the pipeline issue (Dutta, 2004).

Indian Ocean Cooperation and Collaboration

Nevertheless, the trends of cooperation and collaboration between states and global companies have opened up many opportunities in the hydrocarbon sector. The Dolphin project – the world's biggest gas project – was conceived as a joint venture between the United Arab Emirates Offset Groups (UOG), Total of France and

Enron. It was an ambitious scheme to transport natural gas from Qatar's huge offshore North Field via an undersea pipeline to UAE, an example of the emerging dynamics of market integration among neighbouring countries. Through the Dolphin project, Qatar, which is one of the leading producers of natural gas in the Persian Gulf region, was able to provide gas to neighbouring states like Abu Dhabi and Dubai, which were looking for gas to diversify their economies. Today, the Dolphin project is also supplying gas to countries like Oman, Pakistan and India (Doha Butt, 2001).

The other LNG project which has attracted a lot of attention is a deal between China and Iran. Through this deal, the Zhuhai Zhenrong Corporation of China has signed a framework agreement to buy 2.5 million tonnes of super-cooled, compressed natural gas per year from Iran, starting in 2008 and this is expected to rise to five million tones a year from 2013. The Zhenrong Corporation, which is also one of the China's four major state oil traders, had signed an MOU with the Iranian Ministry of Petroleum to undertake development and production of three Iranian oilfields as part of the LNG purchasing plan. It may be mentioned here that there is considerable pressure on Chinese state oil firms to secure foreign oil and gas assets, to fuel its fast growing economy, especially as domestic oil and gas output is declining. China is the world's second largest oil consumer and a net importer of oil. Currently, it is spending billions of dollars on pipelines and LNG ports to boost its natural gas consumption, which is less than 3% of its energy mix now, to 8% by 2010 (The Gulf Today, 2004).

This trend of cooperation and conflict is also reflected in the hydrocarbon sector of India, which is working out an elaborate energy procurement plan through its energy diplomacy, because of the growing demand for oil and natural gas in the country. In the short-run, energy supplies to India would come from Nigeria, Persian Gulf countries and countries in Southeast Asia, particularly Indonesia. In the near future, Bangladesh, Qatar and Turkmenistan are likely to emerge as India's partners for the supply of oil and gas. Oil companies in India have signed a major oil deal with Nigeria.

The participation of ONGC Videsh, Ltd. (OVL) in South Africa's upstream sector has forged energy ties between the two countries (Rajen, 2002). Indian companies are also engaged in oil exploration and production in Central Asia, Africa and the Persian Gulf region, in collaboration with global companies. In fact, India has a visible presence in oilfield operations in 7 countries, extending from eastern Russia to Venezuela. Indian companies like GAIL and OVL have started construction of a trans-Asian oil pipeline from Kazakhstan to the Indian Ocean coast. On the other hand, the consortium of OVL-GAIL-DAEWOO-KOGAS announced the discovery of a giant gas field at Shwe, located in the northwest of Myanmar. The estimated recoverable reserve of the Shwe discovery is in the range of 4 to 6 trillion cubic feet of gas (which is equivalent to 700 million to 1000 million barrels of oil). OVL also invested in an oil field in Sudan. These efforts at co-operation in the hydrocarbon sector by India are seen as building blocks in India's quest to achieve energy security. However, the investment by OVL of India in an oil field in Sudan, made in March 2003, has met with obstacles. It may be recalled that OVL decided to buy a 25% stake in Talisman Energy, a Canadian firm engaged in the Greater Nile Oil Project. It decided to invest US $750 million in the Sudanese oil field and acquired, for the first time, a stake in a foreign oil field. According to the terms of the deal, OGL's share from the oil field comes to 3 million tonnes of oil and gas per annum. The first consignment of crude oil from the oil field was received in May 2003. However, Sudan has been going through a protracted civil war for the last 40 years, between the people in the north, predominantly Arabs Muslims, and Christians and Blacks from the South, resulting in the death of more than 2 million people and the displacement of many more. Owing to pressures from human rights organisations, the US oil company, Chevron, has withdrawn its operation in Sudan since 1984. However, the Sudanese Government renewed its operation in the petroleum sector by striking partnerships with various oil companies in Asia, Europe and Canada. However, the European and Canadian firms involved in oil exploration in Sudan came under severe criticism from Amnesty International and other human rights organisations and church groups. It seems that the ruling National Islamic Front Government (NSF) government of Sudan has been using oil revenues to buy weapons, to suppress the opposition led by the Sudan People's Liberation Army (SPLA). Consequently, the SPLA has dragged India into the civil war, and warned it of dire consequences if it cooperates with the NSF. India's dilemma is that, on the one hand, it has to meet the rising domestic demand for oil and therefore cannot keep away from its involvement in Sudan's oilfield exploration. On the other hand, it cannot overlook the risks involved in engaging in oil exploration in a country dominated by civil war. Nevertheless, India has taken the risk and ignored the rebel threats, as part of its energy procurement plan (Rajen, 2002).

Conclusion

In conclusion, it may be said that there is a need for continuous contact and dialogue among states and global actors, for collaborative efforts in the hydrocarbon sector. It is imperative to continue the process of dialogue among consumers, governments, oil companies and human rights activists, to arrive at a common approach to security questions as well as to regional sensitivities associated with oil exploration. The objective of energy security can be achieved through collaboration in development of technology and its deployment and through joint action by various groups against the threats to those who are putting in efforts to bring about energy security. Here, regional organizations in the Indian Ocean region like IOR-ARC, SAARC and ASEAN can together play a major role in defusing tensions between states, and help to bring about greater understanding and cooperation among them on the key energy security issue for the sake of mutual prosperity.

References

Doha Butt, G., "The Dolphin Project will set a new standard for the Gulf gas sector", www.ameinfo.com

Dutta, S. (2004), "Gas Pipelines are likely to remain as a Pipedream", *The Times of India*, Mumbai, January 13.

International Energy Agency (IEA) (1996), *Asian Gas Study*, International Energy Agency, Paris.

International Energy Agency (IEA) (2001), *World Energy Outlook*, International Energy Agency, Paris.

Leipziger, D and Mudge, J. L. (1976), *Seabed, Mineral Resources and the Economic Interests of Developing Countries*, Cambridge, Massachussets, Ballinger Publishing Company.

Pant, G. (2001), "Globalisation and cooperation-conflict in the Indian Ocean Region", paper presented in an international seminar organised by Centre for African Studies, University of Mumbai.

Rajen, H. (2002), "Recasting Indo-African Development Cooperation," *Economic and Political Weekly*, October 5-11.

SIPRI (1974) Yearbook

The Gulf Today (2004), "China, Iran sign a $20B LNG deal", March 19.

www. earthrights.org/burma

CHAPTER 6

Conflict and Cooperation in Managing Maritime Space in the Persian Gulf: Implications for Energy Security

Vivian Louis Forbes

Introduction

This chapter argues that, in order to maximise energy security in the Indian Ocean Region, it is essential to develop strategies for cooperation in managing maritime space and to develop appropriate legal instruments for the delimitation and administration of the coastal zone. In order to develop this argument, the chapter deals in detail with the Persian Gulf as a case study. Following an overview of the location of the Gulf in relation to global energy security, the chapter then discusses in more detail its geopolitical context. A description and evaluation of Gulf practice in relation to the Law of the Sea Conventions is then presented and this is followed by a detailed discussion of maritime boundary delimitation in the Gulf. In the final section, the potential for cooperation and conflict in the Gulf is considered in the context of maximising energy security.

The Persian Gulf and Global Energy Security

The Persian Gulf (sometimes referred to as the *Arabian Gulf* or the

Gulf) has been in the geopolitical focus since the 1960s. In this chapter, the short-form names for the states will be used throughout and the individual Emirates will be collectively termed the United Arab Emirates. The present author appreciates the sensitivity that is attached to the name of this water body and therefore has chosen to use the most common name as depicted traditionally on most atlases found in secondary and tertiary educational institutions and in public libraries. As a matter of interest, on 24 November 2004, Iran's Culture and Islamic Guidance Ministry banned the sale of a National Geographic Society publication (atlas), which was released in October 2004, to protest the use of the term *"Arabian Gulf"* in the atlas, alongside that of the name 'Persian Gulf' and to delete reference to comment that Iran had "occupied" several Gulf Islands (Associated Press, 24 November 2004). The news item also noted that Hassan Rowhani, Secretary of Iran's Supreme National Security Council, commented that "changing historical names was a dangerous violation of geographical borders and works counter to the stability of the region". Such is the political sensitivity to the naming of geographical features in a region.

The adjacent landmass, the waters and the substratum of the seabed of the Persian Gulf are resource rich, especially with an abundance of hydrocarbon reserves. Indeed, in 2002, according to the US Department of Energy's Administration, the Persian Gulf States (Bahrain, Iran, Iraq, Kuwait, Oman, Qatar, Saudi Arabia and the United Arab Emirates) produced about 25 per cent of the world's oil, while holding nearly 66 per cent of the world's crude oil reserves. The OECD (Organisation for Economic Co-operation and Development) gross oil imports from the Persian Gulf countries averaged about 10.6 million barrels per day during 2002, thereby accounting for nearly 27 per cent of the OECD's total gross oil imports. Besides oil, the Persian Gulf region has huge reserves (1,923 trillion cubic feet) of natural gas, equating to about 36 per cent of total proven world gas reserves. These potential hydrocarbon resources and reserves, the quota, production and capacity of the Organisation of Petroleum Exporting Countries (OPEC) as depicted in Table 6.1, the political alliances, the common religion, albeit factional, and the geostrategic importance of the Gulf are all factors that offer material for academic research and attract electronic and print media attention. Five of the littoral

States in the Persian Gulf are members of OPEC. Iraq and Oman are not affiliated with the Organisation.

The states of the Arabian Peninsula and Iran value their maritime and terrestrial territory highly for its endowment of hydrocarbon reserves (Graz, 1992). The invasion of Kuwait by the forces of former President Saddam Hussein of Iraq in August 1990 marked a major watershed in the modern history of the Middle East region. The demarcation of the boundary with Kuwait by the UN Commission in 1994 was forced on Iraq and thus failed adequately to deal with Iraq's practical need for unfettered access to the sea from its port of Umm Qasr (Muir, 2004:148). The 'invasion of Iraq' by United States forces and its allies in 2003 in their quest to locate weapons of mass destruction and overthrow the then Iraqi Government, led by President Saddam Hussein, marked another turning point in the geopolitics and instability of this troubled region.

Historical issues and contemporary problems that are geopolitical in nature are generally obstacles to dispute resolution in this region (Schofield, 1994). The following three examples briefly illustrate the point. First, a territorial dispute over the Hawar Islands and the precise alignment of a maritime boundary existed between Bahrain and Qatar. On 16 March 2001, the International Court of Justice (ICJ) awarded the Hawar Islands to Bahrain and realigned the maritime boundary between the two states; a final terrestrial boundary resolution was agreed between Saudi Arabia and Qatar in March of 2001. In some instances, talks held in a spirit of brotherhood and understanding have produced agreements— for example, a maritime border pact that was signed on 15 December 2003 by representatives from the Governments of Oman and Yemen. Details of this and many others relating to cooperative ventures are often not placed in the public domain.

Second, an article in the *Gulf News Online* of 12 October 2003, reported that Oman and the UAE exchanged 'authenticated documents,' presumably the instruments of ratification, for the land boundary agreement that was signed on 22 June 2002. The UAE Deputy Prime Minister declared that the exchange brought an end to the process of border demarcation between the two countries, from Umm Zamoul to Al Daar in the north.

Table 6.1. OPEC oil production.
(Thousand of Barrels Per Day)
(Energy Information Administration\Short-Term Energy
Outlook – December 2004)

	11/01/2004	October 2004	November 2004		
	OPEC 10 Quota	Production	Production	Capacity	Surplus capacity
Algeria	862	1,250	1,250	1,250	0
Indonesia	1,399	945	940	940	0
Iran	3,964	3,900	3,900	3,900	0
Kuwait	2,167	2,400	2,400	2,400	0
Libya	1,445	1,560	1,560	1,560	0
Nigeria	2,224	2,300	2,300	2,300	0
Qatar	700	800	800	800	0
Saudi Arabia	8,776	9,500	9,500	10,000-10,500	500-1,000
United Arab Emirates	2,356	2,500	2,500	2,500	0
Venezuela	3,107	2,500	2,500	2,500	0
OPEC 10	27,000	27,655	27,650	28,150-28,650	500-1,000
Iraq		2,200	1,800	1,800	0
Crude oil total		29,855	29,450	29,950-30,450	500-1,000
Other liquids		3,903	3,905		
Total OPEC supply		33,758	33,355		

Notes: Crude oil does not include lease condensate or natural gas liquids. OPEC quotas are based on crude oil production only. "Capacity" refers to maximum sustainable production capacity, defined as the maximum amount of production that: 1) could be brought on-line within a period of 30 days; and 2) sustained for at least 90 days. Kuwaiti and Saudi Arabian figures each include half of the production from the Neutral Zone between the two countries. Saudi Arabian production also includes oil produced from its offshore Abu Safa field produced on behalf of Bahrain. The amount of Saudi Arabian spare capacity that can be brought online is shown as a range, because a short delay may be needed to achieve the higher level. The United Arab Emirates (UAE) is a federation of seven emirates. The UAE's OPEC quota applies only to the emirate of Abu Dhabi, which controls the vast majority of the UAE's economic and resource wealth. Venezuelan capacity and production numbers exclude extra heavy crude oil used to make orimulsion. OPEC: Organisation of Petroleum Exporting Countries: Algeria, Indonesia, Iran, Iraq, Kuwait, Libya, Nigeria, Qatar, Saudi Arabia, the United Arab Emirates, and

Venezuela. OPEC 10 refers to all OPEC less Iraq. Iraqi production and exports have not been a part of any recent OPEC agreements. Iraq's current production number in this table is net of re-injection and water cut. Latest estimated gross production is about 2.3 million barrels per day. Other liquids include lease condensate, natural gas liquids, and other liquids including volume gains from refinery processing.

Source: http://www.eia.doe.gov/emeu/steo/pub/3atab.html (accessed 10 December 2004)

Third, a 20-km neutral zone between Saudi Arabia and Yemen was established in 2000. On 7 February 2003, leaders of the Wayilah tribe issued a statement protesting the Yemeni-Saudi Arabia border committee's Memorandum of Understanding that demanded tribesmen identify their properties outside the international boundary. A Yemeni tribal sheik claimed that Saudi Arabia had built a security fence that was four to seven kilometers beyond the neutral zone inside Yemen, stretching from Jabal Hobash to Jabal Al Fara, and commented that 3,000 tribesmen were ready to fight at any time if Saudi Arabia did not remove the fence (*Yemen Times*, 12 February 2003). Saudi Arabian officials acknowledged the fact and accepted the removal of the fence after extensive diplomatic representation from Egypt and the United States. The demarcation of terrestrial boundaries and the delimitation of maritime boundaries and subsequent delineation of these lines on maps and charts, respectively, could only result in real partnership and to future bilateral and multilateral relationships and cooperative ventures in the regional context.

The Geopolitical Context

The Persian Gulf is a semi-enclosed sea that extends in length for about 455 nautical miles (approximately 840 kilometres) from the coastline of the United Arab Emirates (UAE), the southern sector of the water body, to Al Faw peninsula and the mouth of the Shatt al Arab waterways, shared by Iran and Iraq, in the northwest sector (The Hydrographer, 1967). The width of this semi-enclosed sea varies from about 20 nautical miles (M) at the narrowest, in the northern sector, to nearly 240 M at the widest portion, somewhere in the vicinity of an imaginary line joining the Musandman and Qatar peninsulas. Within the Persian Gulf the depth of water rarely exceeds 50 metres and it increases more rapidly from the shoreline of Iran than from the Arabian coast. The surface area of the Persian Gulf is about 242,000 square kilometres.

Despite a relatively large annual inflow rate of freshwater from the Euphrates and Tigris River system, salinity is high in the Persian Gulf due to the low precipitation and high water evaporation. Highly saline water enters from the Gulf of Oman forming a counter-clockwise gyre then exiting as a submerged, denser, warmer and saline water mass moving towards the centre of the Indian Ocean. Water temperatures in the Persian Gulf show a high seasonal fluctuation. The Persian Gulf is shallow, although deeper nearer the Iranian coast and fringed with extensive coral areas on the Arabian side. Around the Gulf of Oman the continental shelf is extremely narrow, and fisheries concentrate on pelagic species, which, in the last decades, has increasingly meant the medium-sized pelagic such as *Scomberomorus spp* and various species of tuna (FAO, 1997).

There have been significant territorial disputes; and yet, in some select instances, a mutual understanding to push aside historical issues that have a potential for conflict between the Persian Gulf states has been witnessed. For example, the Iran-Iraq War of 1980-1988; the Iraqi Invasion of Kuwait on 1 August 1990; the dispute over sovereignty to three islands – Abu Musa, Greater and Lesser Tunb and Islands between Bahrain and Qatar are cause for many of the geopolitical problems that existed in the region and this semi-enclosed sea.

A Neutral Zone, whose limits were vaguely defined, separates part of Kuwait and Saudi Arabia in the vicinity of Lat. 28° 30′ N. and 48° 00′ E. This was established by the *Kuwait-Nejd Convention* which was signed on 22 December 1922 and is commonly referred to as the '*Uqair Agreement*'. Much later, the terrestrial limits were delimited with some precision for administrative purposes, although the seaward projection of the boundary into the adjacent sea was not delineated. Thus, Kuwait's sovereignty over Qaruh and Umm al Maradim islands was disputed by Saudi Arabia. Furthermore, the rights to explore and exploit the potential resources contained therein became a major issue for a third party, namely Iran.

Iran "occupies" two islands in the Persian Gulf claimed by the United Arab Emirates: Lesser Tunb (called *Tunb as Sughra* in

Arabic by UAE and *Jazireh ye Tonb-e Kuchek* in Persian by Iran) and Greater Tunb (called *Tunb al Kubra* in Arabic by UAE and *Jazireh ye Tonb-e Bozorg* in Persian by Iran); it jointly administers with the UAE an island in the Persian Gulf claimed by the UAE (called *Abu Musa* in Arabic by UAE and *Jazireh ye Abu Musa* in Persian by Iran) over which Iran had taken steps to exert unilateral control since 1992, including access restrictions and a military build-up on the island; the UAE garnered significant diplomatic support in the region in protesting these Iranian actions (Forbes, 1995). After their Eight-year War Iran and Iraq restored diplomatic relations in 1990 but are yet to settle disputes concerning border demarcation, freedom of navigation and sovereignty over the Shatt al Arab waterway.

The demilitarised zone (DMZ) that separates Iraq and Kuwait was established by the U.N. Security Council immediately after the Gulf War in 1992 to keep the armies of the two countries apart and disarmed within a border zone that extends ten kilometres into Iraqi territory and about eight kilometres into Kuwaiti space. Its demarcation is fortified with an electrified fence and concertina-wire barrier. The DMZ runs the full length of the nearly 200-kilometre terrestrial boundary from Saudi Arabia to the Persian Gulf. It is protected by a five-metre deep, five-metre wide and a three-metre high dirt trench. It is a formidable barrier; that was the intention! The barriers to cooperation on land are in stark contrast to the urging of unified management of the marine environment and resources and the need for a comprehensive energy security policy.

The Law of the Sea Conventions and Gulf State Practice

An international convention on the law of the sea is essential for an economic, legal and political order to manage maritime space, its resources and the natural environment within national jurisdictional limits and beyond into the high seas – the marine commons. In particular, the marine biotic and mineral resources contained within the commons are considered to be held for the greater benefit of humankind. International conventions, and in particular, the 1982 *UN Law of the Sea Convention* in its Article 123, urges states bordering semi-enclosed and enclosed seas to cooperate with each other in exercising their rights and performing their duties under the 1992 Convention.

The Gulf States were not party to the 1958 Geneva Conventions despite the fact that some had concerns and reservations about the exploitation of the marine biotic and mineral resources on and under the continental shelf within the Persian Gulf and jurisdictional issues pertaining therein. Of the Gulf States, Bahrain, Iran, Iraq, Kuwait, Oman and the United Arab Emirates were signatories to the Final Act of the 1982 Convention. Table 6.2 illustrates the status of the 1982 Convention in the context of the Gulf States.

Table 6.2. Gulf states and status of the 1982 convention.

State	Date of Signature	Date of Ratification	Agreement – Part XI
Bahrain	10 Dec. 1982	30 May 1985	–
Iran	10 Dec 1982	–	–
Iraq	10 Dec. 1982	30 July 1985	–
Kuwait	10 Dec. 1982	2 May 1986	2 Aug. 2002
Oman	10 Dec. 1982	17 Aug. 1989	26 Feb. 1997
Qatar	–	9 Dec. 2002	9 Dec. 2002
Saudi Arabia	–	24 April 1996	24 April 1996
United Arab Emirates	10 Dec. 1982	–	–

Source: UN Division for Ocean Affairs and Law of the Sea, Chronological List of Ratifications etc., dated 5 January 2005 (http://www.un.orh/ Depts/los/re.ference files/)

Although Iran has not yet ratified the 1982 Convention and supporting Agreement to Part XI of the Convention, it has nevertheless signed, on 17 April 1998, the Agreement for the implementation of the provisions of the Convention relating to the Conservation and management of Straddling Fish Stocks and Highly Migratory Fish stocks. The other Gulf States had not, by January 2005, ratified this Agreement.

Table 6.3 shows the breadth of jurisdictional zones (measured from the appropriate baselines) claimed respectively as territorial sea (TS), contiguous zone (CZ), exclusive economic zone (EEZ) and fishery zone (FZ), where no EEZ is claimed, as being under the state's jurisdiction. The information is compiled from various, sometimes unofficial, sources; the absence of a limit from this list indicates that the information is not in the public domain.

Table 6.3. National Claims to Maritime Jurisdiction (M).

Country	TS	CZ	EEZ	FZ
Bahrain	12	24		
Iran	12	24	200	
Iraq	12			
Kuwait	12			
Oman	12	24	200	
Qatar	12	24		to median
Saudi Arabia	12	18		
UAE	12	24	200	

Source: UK Hydrographic Office HH. 085/012/01.

Maritime Boundaries Delimited in the Gulf

Table 6.4 presents a list of the approximate lengths of the coastline, the area of the natural continental shelf and EEZ of each of the states bordering the Persian Gulf. Continental shelf boundaries (CSB) have been determined for the central portion of the Persian Gulf, in the Strait of Hormuz and the northern waters of the Gulf of Oman. A brief analysis of each of the adjudicated and negotiated boundaries is presented below and illustrated in Figure 6.1. The scale of the map does not permit a more detailed and precise delineation of the boundaries. A comprehensive analysis of each boundary may be found in studies by the Geographer (US State Dept., for example, *Limits* No. 94, 1981; Charney and Alexander 1991; Forbes 1995).

Bahrain/Saudi Arabia: Continental Shelf boundary and Joint Development Zone

An agreement between these two coastal states on 22 February 1958 determined a continental shelf boundary (CSB) and established a special zone wherein exploration and exploitation of hydrocarbons would be undertaken. The operations of such works would be funded by the Saudi Arabian government but the profits gained from the zone would be shared equally by the two states.

The single maritime boundary was determined on the principle of equidistance. Seven points were chosen on the coast of Bahrain and eight points along the Saudi Arabian coast. The geographical coordinates of 11 mid-points of the lines connecting the specified coastal features on the respective coastlines were determined. Geodesic lines link these 11 turning points. The collective length of the 11 lines is 55 M. From turning point 11 the boundary trends in a northeasterly direction to turning point 14. Geodesic lines 12 to 13 and 13 to 14 form the eastern boundary of the common zone of exploitation. The length of the boundary from point 11 to 14 is 45 nm. (Source: UNTS; *Limits* No. 12)

Table 6.4. Coastal length, areas of continental shelf and EEZ.

State	Coastal length (km)	Shelf area to 200m (sq km)	EEZ area to 200nm
Bahrain	60	1,500	1,500
Iran	990	31,200	45,400
Iraq	100	200	200
Kuwait	130	4,100	4,100
Oman	1,000	17,800	163,800
Qatar	200	7,000	7,000
Saudi Arabia	1,330	22,70	54,900
United Arab Emirates	420	17,300	17,300

Abu Dhabi/Qatar: Continental Shelf Boundary

A strict application of the equidistance principle was not evident in the determination of the CSB between Abu Dhabi and Qatar. The overall length of the boundary is 115 M. It comprises four turning points and lies in water depths ranging from 10 to 40 metres. With the exception of a 15 M arc around Dayyinah Island (as a result of a three-M territorial sea radius) favouring Abu Dhabi, the CSB consists of three geodesic lines. The Agreement was first signed on 19 February 1968.

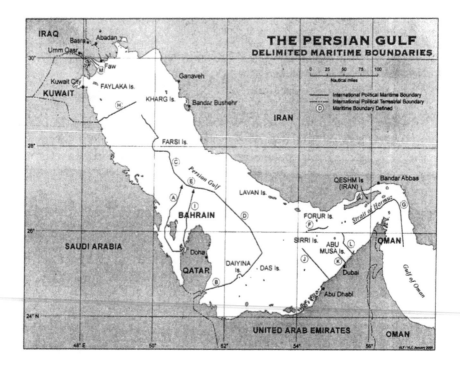

Figure 6.1.

Point A was determined as being apparently the common point of the maritime boundary of three states, being equidistant from the coasts of Abu Dhabi, Iran and Qatar. The line linking point A to point B is 35 M long. Point B was located at the position of an oil well named al Bunduq. Point D, which is the landward terminus point of the boundary, is defined as the intersection of the three-M arcs which represent the limits of the territorial sea of each of the states. The line linking Point D to Point C was delineated so as to be equidistant along selected sectors between the shoals which are part of the respective Abu Dhabi and Qatar sovereignties. (Source: UNTS; *Limits* No. 18)

Iran/Saudi Arabia: Continental Shelf boundary

The 1968 Agreement on the CSB between the two states is a modified version of a median-line negotiation initiated in 1965 but never ratified. Iran's objection to ratification was based on that country's opinion that the agreement did not make provision for an equitable sharing of the seabed resources. Perhaps there was justification in its concern, for new mineral resources were discovered in the northern portion of the negotiated zone in late 1965.

The 1965 version of the boundary was essentially a straight line. The revised boundary of 1968 affected the lines between points 8 and 14. Although the modified boundary crosses and re-crosses the 1965 line, there is not much deviation from the earlier line. Claims to sovereignty over the islands of Farsi and Arabi were an issue in the negotiations. The 1965 agreement granted sovereign rights over Arabi Island to Saudi Arabia and Farsi to Iran. The boundary around the islands of Arabi and Farsi was determined by giving effect to the 12 M territorial sea of the respective islands.

The overall length of the CSB, comprising 16 turning points, is 139 M. It lies in water depths ranging from 40 to 80 metres. Although the CSB is not precisely a median-line in its entirety, it nevertheless, achieves its objective where feasible and represents a compromise for the parties to the agreement. (Source: UNTS; *Limits* No. 24)

Iran/Qatar: Continental Shelf Boundary

The negotiated CSB was determined on the basis of the equidistance principle. The presence of islands between the two states in the Persian Gulf, however, was disregarded for the purpose of delineation of the boundary. The agreement entered into force on 10 May 1970.

At the time of negotiations the location of Point 1 was indefinite. Its location would be dependent on the potential boundary between Bahrain and Qatar. Thus the geographical coordinates of Turning Points 2 to 6 are given in the text of the agreement. Geodesic lines connect these points. The overall length of the boundary is 131 run. Point 2 is located about 31 nm from the Bahrain/Saudi Arabia continental shelf boundary. Point 6 coincides with Point A which marks the seaward extent of the Abu Dhabi/Qatar shelf boundary. (*Limits*, No 25)

Bahrain/Iran: Continental Shelf Boundary

The agreement defining the CSB between Bahrain and Iran entered into force on 14 May 1972. Four turning points connected by geodesic lines define the continental shelf boundary between Bahrain and Iran. The total length of the boundary is 28 M. Water depths along the series of lines range from 60 to 80 metres. Whereas Points 1 and 4 of this agreement were determined by existing continental shelf boundary agreements, Points 2 and 3 are equidistant points. Point 1 of the boundary is coextensive with Point 2 of the Iran-Qatar continental shelf boundary. It is located about 10 M nearer Iran than Bahrain. On an average the turning points are 50 M from Iranian territory and 54 M from Bahrain. (Source: UNTS; *Limits* No. 58)

Iran/United Arab Emirates (Dubai): Continental Shelf Boundary

The agreement establishing a CSB between Iran and Dubai of the United Arab Emirates entered into force on 15 March 1975. Five turning points, connected by geodesic lines for a total distance of 39 M, define the CSB between these states. Water depths along this series of lines are less than 55 metres. The boundary is not determined on the equidistance principle evident from the fact that four of the five turning points are nearer to Sirri Island than to any UAE territory. Indeed, the agreement appears to ignore the presence of the islands. However, between Points 3 and 4 the line follows an arc of 12 M which coincides with the southern territorial sea limit of Iran's Sirri Island. The CSB has been determined without reference to the Iranian islands of Qeys, Forur as well as the disputed island of Abu Musa. (UNTS; *Limits* No. 63)

Iran/Oman: Continental Shelf Boundary

The agreement which established the CSB between Iran and Oman entered into force on 28 May 1975. Turning Point 1 of the CSB between these two states is located in the eastern margin of the Persian Gulf. The delineated boundary extends through the Strait of Hormuz and into the Gulf of Oman. Geographical coordinates for points 1 to 21 inclusive are listed in the text of the agreement. Geodesic lines connect these points. The overall length of the boundary is 125 M. Point 22 is defined as being located on an azimuth of 190° from Point 21 to the intersection of the Oman-Sharjah lateral offshore boundary. This latter boundary has not been established. Point 22, in the Gulf of Oman, remains an undefined mark. (Source: UNTS; *Limits* No. 67).

Kuwait/Saudi Arabia: Submerged Zone Contiguous to the Partitioned Zone

An Agreement signed on 2 July 2000, which entered into force on 30 January 2001, is simple in prose that contains ten Articles and an Annex. Article One of the Agreement defines a set of four geographical coordinates. A suite of four lines commencing from Point G, the land terminal location, connect these geographical coordinates to delineate a maritime boundary that divides the submerged area adjacent to the Neutral Zone of a 1922 Agreement.

The Annex noted that the parties agreed that the natural resources in the submerged area adjacent to the divided zone shall be owned in common and that the two islands – Quarah and Umm al-Maradim – are included in those resources.

The signing of the Agreement of 2 July 2000 between the two Governments resolved a number of on-going issues relating to the Kuwait/Saudi Arabia Offshore Neutral Zone (ONZ) and at the same time provided an avenue for the process of negotiating the delimitation of maritime boundaries between Iran and Kuwait and Iran and the Offshore Neutral Zone.

Significant in this agreement was the statement in Article Eight which stated that the parties would be considered a single negotiating party when negotiations to delimit the eastern boundary of the Offshore Neutral Zone took place with Iran in the future.

Bahrain and Qatar: Delimitation of a Maritime Boundary and Solving Territorial Claims

The Bahrain/Qatar maritime boundary was established by a

judgment of the International Court of Justice (ICJ) that became effective on 16 March 2001. In the *Case*, the ICJ adjudged the alignment of the maritime boundary between the two states and awarded the Hawar Islands to Bahrain and gave those islands full effect for the purpose of constructing equidistant lines to define the turning points in the maritime boundary. The ICJ awarded Zubarah to Qatar. Failure to do so would have created an extremely complex issue as there were several other geographical features in the vicinity – reefs and rocks – that would affect the boundary alignment.

The *Case* was taken to the ICJ in 1991. An extended jurisdictional phase in 1994-95 and the complex merits of the proceedings were contributing factors for the prolonged judgment.

Abu Dhabi/Dubai Offshore Boundary

An administrative delineated line established in 1951 for the purposes of oil concessions between the two states was subsequently adopted by the parties as their maritime boundary in 1965. As a result of a modification of their land frontier, the maritime boundary was realigned about ten kilometers westward of the 1951 line in an agreement that was signed and entered into force on 18 February 1968. Ras Hasain, a geographical coastal feature, is the land/sea interface terminal point. The realignment of the boundary places the Fateh Wells hydrocarbon reserves within Dubai's jurisdiction.

Neither state has defined their territorial sea base points and the parties' normal baselines, due in part to a relatively straight coastline, have not affected the alignment of the boundary, which was negotiated in a just and equitable manner.

Dubai/Sharjah Offshore Boundary

This maritime boundary was determined by an ad hoc arbitrations tribunal on 18 October 1981, together with a resolution of a land boundary dispute which included the territorial issue of where the terrestrial boundary meets the sea. Regrettably, the text of the arbitral award is not in the public domain. It could be inferred that the boundary alignment was determined on the equidistance principle.

Approximately 35 M offshore is the island of Abu Musa whose sovereignty status is disputed by Sharjah (UAE) and Iran and is a subject of an agreement between them.

Sharjah/Umm al Qaywayn Offshore Boundary

The prose employed in the agreement between the two parties to delimit their offshore boundary is simple. For example, the boundary is described as "... a line starting from a point on the coast near the site of the dead well Mirdar Bu Salaf and going out to sea on a bearing (azimuth) of 312°". The agreement, which was signed by the rulers of each state, entered into force in 1964.

A seaward projection of the offshore boundary meets the 12-M radius from Abu Musa Island, which is claimed by both Iran and the UAE.

Iraq/Kuwait Maritime Boundary

The line depicted by the alphabet 'M' in Figure 6.1, is the maritime boundary that separates Iraq and Kuwait which was surveyed and delimited by the UN Iraq-Kuwait Boundary Demarcation Commission (I-KBDC) on 20 May 1993, as a direct result of a UN Security Council Resolution 687. The Commission (I-KBDC) undertook an overall survey of the land and sea boundary, demarcating and delineating where necessary the new boundary alignment which was accepted by the Governments of both countries.

The maritime boundary is a mere 29 M in length and it separates the Iraqi mainland from Kuwait's islands of Bubiyan and Warbah. The geographical coordinates of the turning points and terminal points of the entire international political boundary are listed in the Report. The western portion of the maritime boundary is based on the Low-Water Springs level along the Kuwaiti coast; the central and eastern section of the boundary is defined on the equidistance principle. The delimitation process and report guarantees a non-suspendible right of navigation in the channel through which the boundary passes.

Potential Maritime Boundaries in the Gulf

Maritime boundaries in the process of being negotiated or to be determined are those between Kuwait and Iraq, Iraq and Iran, and Iran and Kuwait at the head of the Persian Gulf. In the south-eastern sector of the Persian Gulf, Iran and individual Emirates of the UAE have yet to finalise segments of their common continental shelf boundary. Delays in negotiations are centred on the issue of sovereignty of some small islands, in particular Abu Musa, that lie nearly equidistant between the coastlines of the two states. The insular geography of the region favours Iran in boundary determination if all the islands it claims were to be considered as points of measurements.

The UAE wanted to refer its dispute with Iran to the International Court of Justice (ICJ), but Iran insists on resolving the issue bilaterally. Several GCC states have attempted to mediate in recent years and Iran-UAE tensions have eased somewhat, but Iran insists that it has sovereignty over the islands and that this issue is not negotiable. The United States generally supports UAE's proposals but takes no position on sovereignty although it is concerned about Iran's military improvements to the islands and the issue of freedom of passage and navigation rights in the vicinity of Abu Musa.

Energy Security: Conflict and Cooperation in Maritime Space

Japan's Arabian Oil Co (AOC), as then manager of offshore operations for both Saudi Arabia and Kuwait, discovered the *Al-Dorra Field* in November 1967 and drilling of six exploratory wells then proved commercial reserves of gas and condensates at the field. Shell Oil Company conducted seismic surveys of *Al-Dorra* for Kuwait after the 1991 Gulf War and concluded that the field contained 5-billion barrels of gas and condensate. The Government of Kuwait set up Kuwait Gulf Oil Co under the auspices of the Kuwait Petroleum Corp to manage all offshore developments following the terminations of both the Saudi and Kuwaiti agreements with AOC.

Kuwait and Saudi Arabia intended to begin natural gas development from their jointly-owned offshore *Al-Dorra Field*, Kuwait Minister of Energy Sheikh Ahmed Fahed al-Sabah noted on 2 October 2003. However, a decision to go ahead with the *Al-*

Dorra Field could have implications for a gas-supply agreement between Kuwait and Qatar. The 25-year gas purchase agreement signed in 2002 stalled due to Saudi Arabia's refusal to grant permission for an undersea pipeline to cross its waters (*Platts* 3 October 2003). The decision to develop the controversial *Al-Dorra Field*, in the Partitioned Neutral Zone (PNZ), introduced implications for relations with Iran, which also claims part of the field and whose maritime border with Kuwait remains to be delimited. Sheikh Ahmed made no mention of whether Kuwait had resolved the maritime delimitation issue with Iran, which was to have included sharing a portion of the Al-Dorra Field.

On 2 October 2003, the National Iranian Oil Co. (NIOC) also offered exploration rights for eight oil blocks in the Persian Gulf for international tender. NIOC noted that geological and seismological surveys of the blocks were positive. At that moment, six international companies from Italy, Brazil and Spain had expressed interest in taking on exploration contracts in Iran on a buyback basis. Iran's *Fars News Agency* quoted an official at NIOC's exploration department as saying that the proposals by the bidders in the tender were studied and the winner of the first block notified in late-October 2003.

Kuwait and Saudi Arabia delimited their sea boundary in July 2000, which resulted in the equal sharing of *Al-Dorra* within the PNZ. Kuwait and Iran have been discussing their offshore boundary since 2000, the same year Iran halted, at the request of Kuwait and Saudi Arabia, a drilling programme it had quietly started in the crest of the *Al-Dorra* field. Kuwait had earlier indicated it would wait for the delimitation negotiations with the Government of Iran to be concluded before starting any developmental work at the field. Kuwait also stated that it would offer a portion of the field to Iran in the agreement. Iranian officials were of the opinion that the remarks by the Kuwaiti Minister were designed to put pressure on Iranian authorities to speed up agreement on the delimitation of the boundary. The two countries have signed preliminary agreements for Iranian gas sales through a proposed pipeline to Kuwait. The Kuwaiti deputy Minister for Energy, Isa al-Oun, believed that the project remained a priority (*AFP* 24 November 2003).

Iran is opposed to Kuwait's proposed development of a disputed gas field until negotiations on the delimitation of the

Iranian/Kuwaiti maritime boundary are completed, an Iranian Foreign Ministry official noted (*FWN* Select 27 October 2003). Iranian Oil Minister, Bijan Namdar Zanganeh, observed that, based on an agreement between the two countries' foreign ministries, neither has the right to start development without the consent of the other party. Recoverable gas reserves at the field are estimated at around 7 trillion cubic feet with a potential production capacity of 600 million cubic feet to 1.5 billion cubic feet a day. State-run NIOC has shrugged off the announcement on the development of *Al-Dorra*, arguing that no international company will get itself involved in a disputed field.

Kuwait could take its dispute with neighboring Iran over an offshore gas field to international arbitration, such as the ICJ which has ruled on similar disputes, if bilateral talks failed. Diplomatic efforts were pursued by the foreign ministry to resolve the border issue. The Iranian Foreign Ministry spokesman described remarks made by the Kuwaiti Energy Minister about the offshore *Arash* oilfield [called Al-Dorra by Kuwait] as astonishing. According to a report received by ISNA, Hamid Reza Asefi was reacting to recent remarks made by the Kuwaiti Energy Minister about the *Arash* oilfield. Dr Asefi noted: "So far, the two countries have had several rounds of negotiations to demarcate the borders of the two countries' continental shelf. These negotiations have so far studied regulations governing the demarcation of the borders of the two countries' continental shelf. The next round of talks will take place soon" (*BBC Monitoring Middle East*, 26 November 2003).

Kuwait, it was stressed, will not exploit the prevailing political situation and the presence of US troops in the region to force a settlement in the *Al-Dorra Gas* Field, which is also shared with Saudi Arabia. The dispute dates back to the 1960s, when Iran and Kuwait each awarded an offshore concession, the first to the former Anglo-Iranian Petroleum Co., which became part of BP, and the latter to Royal Dutch/Shell.

Disruption to the production of oil extraction and exportation is of vital concern to the littoral states of the Persian Gulf. Throughout 2004, media reports have illustrated the vandalism in setting alight oil fields and pipelines in Iraq and attempts made to hijack ships engaged in the transportation of oil and gas and other terrorist activities. How secure are the hydrocarbon reserves and how safe are the ships that work the oil and gas field transport?

Security of Hydrocarbon Resources

On Thursday, 16 December 2004, Osama bin Laden's message, which was contained in an audiotape and directed specifically at the Government of Saudi Arabia, was an attempt to show he is . still a major player in his homeland, despite a security clampdown that has sharply curtailed al-Qaida's field for terrorist operations within Saudi Arabia. The message was issued after Saudi Arabian security forces weakened the insurgency both with arrests and an anti-insurgency campaign that undercut support for al-Qaida militants in that country (*Associated Press*, 19 December 2004). At the same time, however, bin Laden's audiotape followed up on an al-Qaida show of strength in the country. Five militants attacked the U.S. consulate in Riyadh and stormed into the inner courtyard, firing guns, grabbing human shields and killing five people. Four of the attackers were killed and one was wounded in an ensuing battle with Saudi forces. No Americans were killed.

A few days later, an address by the Saudi Arabian branch of al-Qaida called for attacks against the oil infrastructure in the Persian Gulf. The statement called on "all mujahedeen . . . in the Arabian Perrinsula" to target "the oil resources that do not serve the nation of Islam". The statement urged al-Qaida members and sympathizers around the Arab world to unite "to strike all the foreign targets in the Arabian peninsula and attack all the infidels' havens everywhere".

Analysts described the bin Laden's message, which included a call to followers to "concentrate your operations" on oil facilities, as a reminder that he can still cause trouble. Some also saw it as a sign that bin Laden is worried about effects to his credibility or that he might lose more influence if local elections prove a success. "It's a kind of encouragement for Saudis influenced by him after the blow they have received," said Dia'a Rashwan, an Egyptian expert on militant groups. "It's a way for him to tell them he supports them, he cares about them and will, through his statement, put their cause in the international spotlight" (*Associated Press*, 19 December 2004). In the audiotape, bin Laden exonerated Islamic militants of responsibility for the violence in the kingdom, saying it was the rulers' "sins which exposed the country to God's punishment". He also reiterated long-standing accusations that the royal Al Saud family had misused public funds and allied itself with the "infidel America against Muslims". Addressing the Saudi rulers, bin Laden said: "You must know that people are fed up . . . security will not be able to stop them". Osama bin Laden's direct focus on the Kingdom and its rulers is the first in about a decade. He embarked on his terrorist path in the early-1990s, after the ruling family turned down his request to use his 'mujahedeen' holy warriors, who had trained with him in Afghanistan to liberate Kuwait from Iraqi occupation.

Some of the terrorist activities in Iraq during 2004 and early-2005 allegedly have the financial and moral backing of the al-Qaida group. This and other terrorist groups have realised the importance of global maritime trade, and especially the transportation of petroleum products and Liquified Natural Gas, and could launch attacks against the merchant ships and coastal zone facilities.

Conclusion

In order to develop strategies for cooperation in managing maritime space in the Persian Gulf, and thus enhance regional and international energy security, appropriate national legal instruments for the administration of the coastal zone are factors that must be considered. The Persian Gulf is a special area that requires exceptional protection. Maritime boundary determination, generally Continental Shelf Boundaries, for the central portion of the Persian Gulf, in the Strait of Hormuz and the northern waters of the Gulf of Oman are established. Two sectors of the Gulf – in the north and southeast – have yet to be delimited. These will be achieved when the myriad geopolitical and historical issues within the Persian Gulf are resolved either through bilateral talks or via the intervention of an adjudicator. Large expanses of un-demarcated terrestrial boundaries and the alignment of their seaward projection often lead to disputes. A lot of this will be achieved when the states of the region demonstrate a genuine desire to work together. Threats from militant groups must be counteracted in advance of potential and serious damage being sustained.

Economic issues are most likely to be the motivating factors to get the parties to the negotiating table. The need to ensure the security of offshore installations, the safe passage of the more than 30,000 ships that transit the waters, and the protection of fisheries in this semi-enclosed sea all require a high level of coordinated maritime patrols and surveillance. Oil and gas and related by-products are the entire economic base of many of the littoral states of the Persian Gulf, and even those who might be critical of the West or the United States will be greatly offended by any destruction of their main means of economic progress and survival. It would be hard for the militants to carry out a huge attack on oil installations in the Kingdom that would significantly disrupt production or distribution. At most, isolated attacks could be launched on less-guarded areas and it may be that targeting the kingdom's oil industry could actually backfire.

References

Associated Press, 24 November 2004, 'Iran bans National Geographic Atlas', Media report

Associated Press, 19 December 2004, 'Bin Laden said to aim message at Saudis', Media report

AFP 24/11/2003 'The Al-Dorra Field a Priority', Media report

Al Mazidi, Feisal (1993) The Future of the Gulf, The legacy of the war and the challenges of the 1990s, London: I.B.Tauris

BBC Monitoring Middle East, 26 November 2003 'Negotiations to determine boundary'

Charney, J.I. and Alexander, L.M (eds.) International Maritime Boundaries, The Hague: Martinus Nijhoff Publishers. Four Volumes

Energy Information Administration, 2003 Persian Gulf Oil and Gas Exports Fact Sheet, Country Analysis Briefs (PDF version) April 2003. (eia.doe.gov Website)

FAO (Food and Agriculture Organisation) 1997 Status of World Fisheries, Rome: FAO

Forbes, V.L. (1995) The Maritime Boundaries of the Indian Ocean Region, Singapore: Singapore University Press. 267pp

Forbes, V.L. (2001). Conflict and Cooperation in Managing maritime Space in Semi-enclosed Seas, Singapore: Singapore University Press. 384pp

Forbes, V.L. (1980-2000) The Indian Ocean Newsletter and The Indian Ocean Review. A regular series of articles on maritime boundary determination and maritime jurisdiction appear in these journals.

FWN Select 27/10/2003 'Iran opposed to Kuwait's proposal', Media report

Graz, Liesl (1992) The Turbulent Gulf People, Politics and Power, London:

I.B. Tauris & Co., Ltd.

Gulf News Online 12 October 2003 'Oman and UAE exchange documents'(www.gulf-news.com)

Hydrographer (The), 1967 Persian Gulf Pilot, 11th Edition, London: Hydrographic Department.

Muir, Richard (2004) 'The Iraq-Kuwait Border Dispute: Still a Factor for Instability', *Asian Affairs*, Vol. XXXV, No. 2, 147-161.

New York Post 3 January 2004 'US Naval Forces in Gulf seize small boat carrying drugs', Report

Oman Observer 2003 `Oman, Yemen sign maritime border pact', (Web page: omanobserver.com)

Platts 3/10/03 'Development of the *Al-Dorrra* Field'

Schofield, Richard (Ed.) 1994 *Territorial Foundations of the Gulf States*, London: UCL Press

United Kingdom, Hydrographic Office 2001 *Annual Notice to Mariners*, HH 085/012/01

United Nations Division of Ocean Affairs and Law of the Sea, Text and Status of Treaty http://www.un.org/Depts/los/los94st.htm

UNEP (1991) Overview on Land-based Sources and Activities Affecting the Marine Environment in the ROPME Sea Area, UNEP Coordination Office and ROPME, 127 pp.

UNEP (1995) UNEP *Global Environment Outlook* (http://www1.unep.org/geo-text)

UNEP (1999) *UNEP Global Environment Outlook* (as above)

UNEP (2000) UNEP *Global Environmental Outlook* (as above)

UNTS (United Nations Treaty Series) Various issues listing bilateral treaties, NY: United Nations

United States, Limits refers to the *Limit in the Sea* Series prepared by the Geographer, United States, for example, Limits No. 25, 1970 (case studies of negotiated boundaries).

United States Department of Energy (Webpage http://www.eia.doe.gov/emeu/steo/pub/3atab.html)

Yemen Times, 12 Feb, 2003 'Yemeni tribesmen ready to fight for removal of fence'

ANNEXE I

INTERNATIONAL AGREEMENTS CITED: DETERMINATION OF MARITIME BOUNDARIES

Agreement concerning the delimitation of the continental shelf between Saudi Arabia and Bahrain, done on 22 February 1958, signed and ratified 26 February 1958

Offshore Boundary Agreement between the Emirates of Abu Dhabi and

Dubai, signed 18 February 1968.

Abu Dhabi-Qatar Continental Shelf Boundary, 1969

Agreement concerning the Sovereignty over the Islands of al-'Arabiyah and Farsi and the   Delimitation of the Boundary Line separating the Sub-marine Areas between the Kingdom of Saudi Arabia and Iran Done at Tehran on 24 October 1968 and ratified on 29 January 1969.

Agreement for Settlement of the Offshore Boundary and Ownership of Islands between Abu Dhabi and Qatar, signed and ratified on 20 March 1969

Agreement of settlement of maritime boundary lines and sovereign rights over islands between Qatar and Abu Dhabi (20 Mar 1969)

Agreement concerning the boundary line dividing the continental shelf between Iran and Qatar (20 Sep 1969).

Agreement concerning the Boundary Line dividing the Continental Shelf between Iran and Qatar; signed in Doha on 20 September 1969 and ratified on 10 May 1970

.greement concerning Delimitation of the Continental Shelf between Bahrain and Iran, signed at Bahrain on 17 June 1971, and entered into force on 14 May 1972.

Agreement between Sudan and Saudi Arabia relating to the joint exploitation of the natural   resources of the sea-bed and the sub-soil of the Red Sea in the Common Zone (16 May 1974) Agreement concerning the boundary line dividing parts of the continental shelf between Iran and the United Arab Emirates (13 Aug 1974)

Iranian-Oman Joint Patrol of the Strait of Hormuz, 1974-1977

Agreement concerning Delimitation of the Continental Shelf between Iran and Oman; signed on 25   July 1975 and ratified on 28 May 1975.

Agreement concerning delimitation of the continental shelf between Iran and Oman. (25 Jul 1974)

Treaty Relating to the State Boundary and Good Neighbourliness between Iran and Iraq, 1975

Protocol Relating to the Delimitation of the River Boundary between Iran and Iraq, 1975

Protocol Relating to the Re-demarcation of the Land Boundary between Iran and Iraq, 1975

Protocol Relating to the Security of the Boundary between Iran and Iraq, 1975

Kuwait Regional Convention for Co-operation on the Protection of the Marine Environment from Pollution, 1978

Protocol Concerning Regional Co-operation in Combating Pollution by Oil and Other Harmful   Substances in Cases of Emergency, 1978

Agreement concerning the Boundary Line dividing Parts of the Continental Shelf between Iran and the United Arab Emirates

(Dubai), signed on 13 August 1974.
Understanding of Memorandum concerning the Island of Abu Musa

ANNEX II
NATIONAL LEGISLATION CITED

Bahrain

Proclamation with respect to the sea bed and the subsoil of the high seas of the Persian Gulf, 5 June 1949

Iran

Act on the exploration and exploitation of the natural resources of the Continental Shelf of Iran, 19 June 1955. Act amending the Act of 15 July 1934 on the territorial waters and the contiguous zone of Iran, 12 April 1959 Proclamation concerning the outer limit of the exclusive fishing zone of Iran in the Persian Gulf and the Sea of Oman, 30 October 1973

Iraq

Official proclamation concerning the Continental Shelf of Iraq, 23 November 1957 Supplementary Proclamation concerning the Continental Shelf of Iraq, 10 April 1958 Law No. 71 of 1958 delimiting the Iraqi territorial waters

Kuwait

Proclamation of the Shaikh of Kuwait with respect to the sea bed and the subsoil of the high seas of the Persian Gulf, 12 June 1949 Decree regarding the delimitation of the breadth of the territorial sea of the State of Kuwait, 17 December 1967

Oman

Decree concerning the territorial sea, continental shelf and exclusive fishing zones of the Sultanate of Oman, 17 July 1972 Decree No. 44 of 15 June 1977 (amending art. 6 of the Decree of 17 July 1972)

Qatar

Proclamation with respect to the sea-bed and the subsoil of the high seas of the Persian Gulf, 8 June 1949 Declaration concerning the exclusive sovereign rights of the State of Qatar in the zones contiguous to the territorial sea, 2 June 1974

Saudi Arabia

Royal Pronouncement concerning the policy of the Kingdom of Saudi Arabia with respect to the subsoil and sea bed of areas in the Persian Gulf contiguous to the coasts of the Kingdom of Saudi Arabia, 28 May 1949 Royal Decree concerning the territorial waters of the Kingdom of Saudi Arabia, No. 33, 16 February 1958 Saudi Royal Decree relating to the acquisition of the Red Sea Resources No. M/27, dated 9-7-1388 (1 October 1968)

Declaration concerning the limit of the exclusive fishing zones of Saudi Arabia in the Red Sea and the Arabian Gulf, 1974

United Arab Emirates

Abu Dhabi – Proclamation with respect to the sea bed and subsoil of the Persian Gulf, 10 June 1949.

Ajman – Proclamation with respect to the sea bed and the sub soil of the high seas of the Persian Gulf, 20 June 1949

Dubai – Proclamation with respect to the sea bed and the sub soil of the high seas of the Persian Gulf, 14 June 1949

Ras al-Khaymah – Proclamation with respect to the sea bed and the sub soil of the high seas of the Persian Gulf, 17 June 1949

Sharjah – Proclamation with respect to the sea bed and the sub soil of the high seas of the Persian Gulf, 16 June 1949

– A Decree concerning territorial waters of Sharjah Emirate and its Dependencies and its Islands, 10 September 1969.

– Supplementary Decree concerning the territorial sea of the Emirate of Sharjah and its Dependencies, 5 April 1970.

STAKEHOLDERS IN INDIAN OCEAN ENERGY SECURITY

CHAPTER 7

India's Quest for Energy Security: The Geopolitics and the Geoeconomics of Pipelines?

Sanjay Chaturvedi

Public debate may still be hostage to the outdated vocabulary of political borders, but the daily realities facing most people in the developed and developing worlds, both as citizens and consumers—speak a vastly different idiom. Theirs is a language of an increasingly borderless economy, a true marketplace. But the references we have –the maps and the guides—to this new terrain are still largely drawn in political terms (. . . and) in a borderless economy; the nation-focused maps we typically use to make sense of economic activity are woefully misleading. We must, managers and policy-makers alike, face up at last to the awkward and uncomfortable truth: the old cartography no longer works.

Kenichi Ohmae 1995

If the players left in the field by the waning importance of military power were purely economic entities - labour sellers, entrepreneurs, corporations — then only the logic of commerce would govern world affairs...the action on all sides would unfold without regard for frontiers . . . But things are not quite that simple . . . As territorial entities, spatially rather than functionally defined, states can not follow a commercial logic that would ignore their own boundaries. As bureaucracies write large, states are themselves impelled by the bureaucratic urges of self-preservation and role enhancement to acquire a 'geo-economic' substitute for their decaying geopolitical role.

Edward Luttwak 1990

If it is politically difficult for India and Pakistan to have a bilateral agreement for the joint import of gas via a pipeline, it might be easier to have agreement for a Natural gas transmission system linking several countries. Such a system could extend from the Gulf States, Saudi Arabia, and Iran in the West, Turkmenistan and Afghanistan in the North, through Pakistan and India to Bangladesh. Natural gas could be fed into the pipeline by various countries, base on long-term Agreements, and withdrawn initially by Pakistan and India. The pipeline system could be extended later to Myanmar and Thailand in the East, and to other Central Asian countries in the North.

Toufiq A. Siddiqi 2003

Introduction

As the year 2005 unfolds, one finds a noticeable improvement in India-Pakistan relations. A series of initiatives, aiming at 'crossing' the physical as well as mental borders, has recently been introduced. For example, a two-Punjab Centre has been set up at Chandigarh with considerable diplomatic support from both countries. One of the key objectives of this Centre is to initiate and promote wide-ranging cooperation between the Pakistani Punjab and the Indian Punjab. There is considerable optimism on both sides of the border that one of the major breakthroughs that we might witness, sooner than later, between the two estranged neighbours, could be the laying down of a 'peace' pipeline, originating in southern Iran, traversing the territory of Pakistan and entering the sovereign realm of India. The Indian cabinet accorded its approval to talks on gas pipelines on the 9th February 2005 and soon thereafter the commerce secretaries of both India and Pakistan were being authorized to lead their respective delegations for the trade expert group meetings. An important task before the joint expert-group would be to identify bottlenecks in bilateral trade (which, officially speaking, at present is only half a billion dollars) and how unwanted trade barriers can be removed by either side.

The government of India, demonstrating unprecedented sensitivity to the country's critical requirement for energy resources, has also decided to drop its insistence on linking progress on the proposed Iran-Pakistan-India gas pipeline with trade and transit concessions from Pakistan. The Petroleum Ministry has also been authorised by the Indian Cabinet to proceed with proposals for pipelines to India from Iran, from Turkmenistan through Afghanistan and Pakistan and also from Myanmar through Bangladesh. India's External Affairs Minister, Mr. Natwar Singh, has recently stated in Islamabad, "We have now agreed to consider a pipeline through Pakistan subject to satisfaction of our concerns related to security and assured supplies". The Pakistani Prime Minister, Mr. Shaukat Aziz, is also reported to have said that: "the peace pipeline diplomacy will benefit all the countries in the region . . . the pipeline would benefit all the parties involved [Iran, Pakistan, India] diplomatically and geopolitically" (*The Dawn*, 25 February 2005).

On 6 June 2005, during the visit of Mr. Mani Shankar Aiyar, India's Petroleum Minister, to Pakistan, both countries decided to set up a joint working group to study the legal, technical, commercial and financial parameters of the Iran-India gas pipeline, aimed at getting the project off the ground by the end of the year. Both countries have also agreed to move forward on the Turkmenistan-Afghanistan-Pakistan pipeline to make a TAPI (Turkmenistan-Afghanistan-Pakistan-India) pipeline besides exploring energy cooperation in the Qatar-Pakistan pipeline. Mr. Aiyar is reported to have said that both the countries have agreed that with the projected growth in energy demand, India and Pakistan would require 500 million standard cubic metre per day (MSCMPD) by 2005. "It implies that over the next five years both of us will not only require gas through the Iran-Pakistan-India gas pipeline but also other available resources, including gas from Turkmenistan, Qatar and stretching to Russia, Saudi Arabia and UAE besides LNG from other countries" ·(*The Tribune*,

Chandigarh, 7 June 2005). Mr. Aiyar is also reported to have appreciated the willingness of Pakistan government to consider the Iran-India pipeline as a stand alone project, integral to the widening economic and commercial cooperation without any links to the ongoing peace talks on the Kashmir issue.

A key objective of this chapter is to critically examine the problems and prospects of India's quest for energy against the backdrop of an increasingly complex interplay between geoeconomics and geopolitics. It will explore, in the first instance, the extent to which the geoeconomics of energy (in)security is compelling India to reorient itself vis-à-vis its immediate neighbourhood and even beyond towards Iran and the Central Asian Republics. This is followed by an investigation into how such options are facilitated and/or constrained by geopolitical factors and forces. The proverbial, billion dollar question here is whether there is any evidence to suggest that geoeconomics is replacing geopolitics, with regard to an increasingly critical domain of energy security, in South Asia. If yes, with what measure of success? If not, why not?

The prospects outlined above by Toufiq A. Siddiqi (2003) depend largely, but not entirely, on the extent to which the political elites in South Asia are both willing and able to introspect and reorient their respective state-centric as well as nationalising geopolitics in the context of the broadening as well as deepening of economic interdependencies. This requires in turn a serious and systematic rethinking of the most classically Cartesian and cartographic of orthodox geopolitical categories: the border. A meaningful quest for energy security must grapple, on the one hand, with the complexities of the interplay between geopolitics and geoeconomics in an era of globalisation. On the other hand lies the challenge of resolving the tension between a state-centric, instrumentalist connotation of geoeconomics (see Luttwak 1990) and a de-territorialisation of state sovereignty and the inadequacy of modern cartography to capture the transformation (see Ohmae 1995).

According to Lacoste (2000), a geopolitical analysis focuses the rivalries between political forces not only in respect of ideology and economic competition, but also territory. By territorial rivalry, Lacoste means the strivings by an actor or an agency for control of a territory in order to exert power and influence over the resources and human inhabitants. The territory under question could be an object of dispute for its *strategic* or *symbolic* meaning. Or it could simply be the field of confrontation between rival forces. The definition of territorial rivalry offered by Lacoste could be usefully modified "by extending the term territory to encompass maritime and airspace elements, since in today's world, rivalries between two antagonists is rarely confined to territory in the sense of onshore territory. Natural resources are also located in the sea, and acquiring control over maritime zones and airspace might be indispensable for exerting power over a targeted piece of territory" (Peters, 1999: 30-31).

Needless to say perhaps, such geopolitical rivalries can exist between states, between regional groupings of states, and between ethnic groups within states. Thus, in short, 'geopolitics' could be defined as power rivalries between different types of power authorities for ideological and economic dominance as well as for control and domination of territory, including maritime zones and airspace. Or, as Michael Klare (2003) puts it, geopolitics or geopolitical competition implies, "contention between great powers and aspiring great powers for control over territory, resources, and important geographical positions, such as ports and harbours, canals, river systems, oases, and other sources of wealth and influence. If you look back, you will find that this kind of contestation has been the driving force in world politics and especially world conflict in much of the past few centuries". As far as geo-strategy is concerned, as Zbigniew Brzezinski (1997) puts it, it implies, "the strategic management of geopolitical interests".

Geoeconomics, according to Edward Luttwak (1990), consists of a series of intentional practices either deployed by 'the state' or through 'the state' by national economic elites in order to win access to free markets, block access to domestic markets, create the basis for economic expansion through state-led investment, training, regulatory exchange, or even through espionage involving industrial and technological secrets. According to Luttwak, the overall effect of the bureaucratic impulses of the intellectuals and institutions of statecraft to find new geoeconomic roles, and of geoeconomic manipulations by interest groups, will vary from country to country and from case to case. But fundamentally, states will tend to act geoeconomically simply because of what they are: territorially-defined entities precisely to outdo each other on the world scene. Whereas Ohmae chooses to disagree with Luttwak and argues instead, albeit in the neoliberal extreme, that such state-centrism is becoming outdated at a practical level by the increasing global organization of capitalism where it is individual corporations and consumers, not states, governments and democracies, that increasingly make the important decisions. Luttwak, according to this perspective, appears to be a prisoner of what Ohmae calls the old cartography, a cartography of states and fixed frontiers that cannot map the globe-girdling network of corporations, trade and communications infrastructure, including oil and gas pipelines.

The issues raised by the debate between Luttwak and Ohmae compel us to ask in the case of India's quest for energy security whether geopolitics is being replaced by geoeconomics. One could also wonder whether the two agendas – realist and liberal – are as mutually exclusive as they appear to be. Luttwak might be state-centric in his analysis but he is still "candid and critical about the kinds of extreme free-market posturing that, like Ohmae's arguments, ignores the continuing role of state practices in shaping globe-girdling economic affairs" (Sparke, 1998: 67). Moreover, could it be that the theorists of geoeconomics as state-led conflict underplay the manner in which some of the industrialized states, especially the United States since the 1980s, have become trans-nationalized? But what about India? Is there any evidence to suggest that, in the case of India, state-centrism is becoming outdated for all practical purposes in the wake of corporate globalisation; where it is individual corporations and consumers, and not intellectuals/institutions of statecraft, that take important decisions?

Before we attempt an answer to the questions raised above, we need to take into consideration the three propositions raised by Susan Strange (1994). First, many seemingly 'autonomous' developments in the world of (geo)politics and (geo)economics have common roots. This could be the result in large part of the similar structural changes in the world economy and society. Secondly, partly due to similar structural changes, a fundamental change seems to have occurred in the nature of diplomacy. The universe of bargaining today is not simply 'international', as governments must now bargain not only with other governments but also with firms and enterprises. The third proposition follows from the second and relates to the significance of firms as actors influencing the future course of transnational relations – not least for the study of international relations and political economy.

The Geoeconomics of Energy Flows in South Asia: Imperatives of Demands and Supply

It is a truism that ensuring an adequate supply of energy at reasonable cost is a major prerequisite for both developed and developing countries, especially in the wake of challenges posed by corporate globalization and the underlying current of industrialization and urbanization. It has been pointed out by a number of experts and analysts that the impact of growing dependence on Middle East oil, on global oil supplies, the price of oil and global oil security on global geopolitics is going to remain profound for many more decades to follow. It has been argued, for example, that, "two very powerful geopolitical factors will decisively determine whether the quest for Middle East oil (mainly Gulf oil) could enhance global oil security or could lead to oil supply disruptions and also instability and conflict in the Asia-Pacific region. The two factors are the United States' growing dependence on oil imports from the Middle East and the Asia-Pacific region's thirst for oil and increasingly likely Chinese dependence on oil from the region" (Salameh, 2003: 1085).

It is important to note that energy in the United States, like the rest of the industrialized world, still comes mainly from fossil fuels, accounting for as much as 90% of the total US primary energy needs during the year 2000. According to the estimates offered by the US Energy Information Administration (EIA), US demand for natural gas will rise by 62%, electricity by 45% and oil by 33% over the next twenty years (see Salameh, 2003, 135). Natural gas, being the cleanest fossil fuel, has become the fuel choice for nearly all of the new electricity-generating capacity in the United States and the rest of the world.

It is equally well established that the proven natural gas reserves of India and Pakistan are so small in comparison to the projected demand for the fuel energy and the economic benefits so substantial that, regardless of the tone of tenor of their bilateral relations, it is vital to construct pipelines from the energy-exporting countries located to their North and West – namely

Turkmenistan, Iran, Qatar, the United Arab Emirates, and Oman –
and Bangladesh to the East. The natural gas is best suited to
provide India and Pakistan with the clean energy they need for
ecologically sustainable development during the first half of the
21st century and beyond.

Various aspects of the geoeconomic rationale of building
common pipelines have been meticulously examined in an
excellent recent study by Toufiq A. Siddiqi (2003). It should suffice
therefore to summarise and highlight a few points rather than
provide a detailed account. As pointed out by Siddiqi (Ibid.), at
current rates of production, India's proven reserves are likely to
last for about 25 years, and Pakistan's for about 36 years.
According to Pakistan's Ministry of Petroleum and Natural
Resources (2000), the estimated demand for natural gas in
Pakistan already exceeds the supply by about 250 million cubic
feet per day (MMcfd). By 2010, this figure is likely to double,
notwithstanding the assumed increased domestic production from
fields discovered in recent years (which are likely to add about 950
MMcfd), and the supply-demand gap is expected to be met by
imports of natural gas. Unless major new domestic reserves are
discovered, imports of natural gas could reach a billion cubic feet
per day (1,000 MMcfd), in view of the ever-increasing demand for
natural gas in all sectors (Ibid.). India is also unable to meet the
demand for natural gas from its domestic sources. Demand for
natural gas in India, mainly from new power generation projects,
fertilizer plants, and industrial users, is projected to soar in the
world's second most populous state. Its insignificant domestic
output of natural gas necessitates that India must import natural
gas to meet its expected explosive growth in demand for the fuel.
An expert group was established by the Ministry of petroleum
and natural gas in 1994 and its projections for natural gas demand
and domestic supply up to the year 2009-2010 are as follows.
Currently, oil and natural gas comprise around 40% of the total
energy needs of the country, and over the next two decades, this is
likely to be around 45%. Given that India's domestic oil
production has remained stagnant at less than 32 million tonnes,
unless massive investments and technology are introduced, it is
unlikely that production will increase substantially. As a result,
the gap between demand and supply will increase over the years
as consumption of oil and natural gas goes up. This high level of

energy imports will lead to a major drain on the country's foreign exchange reserves. Already, India imports around 70% of its total oil demand. Since the early 1990s, India has begun to take up the issue of energy security seriously. There appears to be a growing consensus both among scholars and policy-makers that an energy policy will have to be introduced sooner than later in order to balance the country's growing need for energy without increasing its vulnerability, and, at the same time, cater to the country's environmental well-being (Singh, 2001).

By the first half of the 21st century, India is likely to be among the top four consumers of energy, just behind the US, China and Japan, and ahead of countries like France and UK India's GDP growth rate averaged 6.5% during 1992-97 and despite decreasing slightly to 5% in 1997-98, it picked up again to around 6% in 1998-99 and is expected to average between 6-7% thereafter for the rest of the Ninth five-year plan period. According to government statistics, indigenous production of crude oil is stagnating at around 33 million metric tonnes (MMT) per annum, while demand has grown by over 100 MMT. The country has attracted over $8 billion investment in oil exploration and hopes to attract foreign oil majors in new 20 oil fields, currently on offer under the New Exploration Licensing policy (NELP). During 2003-4, the Oil and Natural Gas Commission (ONGC), Oil India and other companies have made 14 hydrocarbon finds worth over 350 MMT reserves (Kumar 2005). The ONGC is currently negotiating with Russia's Gazprom over a string of major oil and gas deals that could see the cash-rich, zero-debt ONGC Videsh (ONGC's overseas arm) invest up to $20 billion in the next few years (*The Financial Express*, 23 February 2005).

As in most countries, India's Industrial sector is the largest consumer of energy, followed by the transport sector (Dadwal, 2002). According to Siddiqi (2003), "There is likely to be a gap of about 4,000 MMcfd between the supply and demand for natural gas in India by 2004-5, and of 7,400 MMcdf by 2009-10. Even if, as planned, coal bed methane is developed, its contribution will be less than 500 MMcfd. The remaining demand would have to be met by imports, or go unfulfilled."

It is to state the obvious that both India and Pakistan rank very low in terms of human resource development and human security. For a vast majority of 1.3 billion plus people living in

South Asia, conflicts, or incurring expenditure in preparing for conflicts [both India and Pakistan are reported to be spending more than one million US dollars a day on the Siachin glacier conflict], are a major drain on scarce resources that are needed to provide adequate water, shelter, education, health care, and infrastructure needed for economic development. The question is not whether or not India needs pipelines, but when and from where? Who gets, what, when, where and how from these strategic energy resources?

Alternate Routes and Possible Energy Sources for India: Realities, Perceptions and Policies

India with an import bill of over 20 billion dollars and an annual kitty of five to ten billion dollar investment in other countries, is now seeking to access new sources of crude oil, spanning across the Central Asian republics, Africa, Latin America and the 'backyard' of South Asia, involving Bangladesh, Myanmar and other countries. India's new energy policy has three major objectives.

First, India would like to ensure long-term stability in the supply of energy since the country's dependence on oil imports is likely to increase from the present level of 70 per cent to over 85 per cent by 2020. Second, the country is planning to build up an energy grid in the region comprising India, Myanmar, Bangladesh, Pakistan, Nepal, Bhutan and Sri Lanka to ensure a two-way flow of power, natural gas and oil. The strategy is to offer economic benefits and open a gateway for trade with ASEAN countries, hoping at the same time that cooperation among South Asian states is likely to create a congenial environment to resolve their (geo)political disputes. Thirdly, by encouraging the Indian public sector oil companies to acquire stakes in the foreign oil fields, with the help of diplomatic missions, an attempt is to be made to ensure a long-term sustainable supply of fuel at a 'reasonable and stable price' till alternative sources of energy are discovered and meaningfully harnessed. One remarkable feature of the new energy security policy of India is the emergence of 'economic diplomacy' and its growing acceptance by actors and agencies across the 'ideological' spectrum. Over the past couple of years, the Ministry of External

Affairs has offered its 'economic diplomacy' to other relevant ministries, especially the Ministry of Petroleum, through inputs such as information received from more than forty countries on aspects such as strategic reserves, infrastructure, supply lines, and business opportunities being forged with third countries (Bansal, 2002).

From the point of view of geographical proximity and the availability of large proven natural gas reserves, the most likely sources of gas for Pakistan and Northwest India, as stated above, are Iran, Qatar, the United Arab Emirates, and Oman in the west, and Turkmenistan in central Asia. It is amply revealed by preliminary assessments of possible routes and technical feasibility studies undertaken by a consortia of companies that there are no major 'technical' obstacles to building natural gas pipelines from any of these countries to supply the quantities required by India and Pakistan. The proposed pipelines are mainly land-based ones, or ones in shallow waters offshore. An earlier proposal to build a pipeline going directly from Oman to Western India was put on the back burner, due to the technical and economic problems of going through stretches where the water is very deep. India has commissioned a study to re-examine the deep ocean option again, due to concerns about the security of supplies coming through Pakistan (Siddiqi, 2003). I will return to this point later in the Chapter.

Discussions on imports from several countries in the Middle East and Turkmenistan have been held separately by India and Pakistan during the past decade, and memoranda of understanding have been signed with several of them. India had signed an initial MOU with Iran in July 1993, and another in November 1993. An agreement in principle to import natural gas via an offshore pipeline was signed between Oman and India in September 1994.

Substantial work has already been done on the feasibility of the Iran-Pakistan pipeline, and its possible extension to India. In 1994, a report was prepared by the joint working committee of professionals from Iran and Pakistan. It was proposed that a pre-feasibility study needed to be carried out. As many as six multinational energy companies were invited to undertake the - BHP Australia, British Gas, Gaz de France, Petronas Carigali Malaysia, Novacorp Canada, and Shell International. The recommendations submitted by BHP in 1997 indicated that the most economical route would be from Bandar Abbas (Assaluyeh gas field) in Iran to Multan in Pakistan, with a branch line from near Khuzdar to Karachi. The other five companies formed a consortium to explore the feasibility option together. "Their preferred route runs from the giant South Pars field offshore Iran to Pakistan. Either of the proposed routes could be extended to India" (Siddiqi, 2003, 34).

Paul Samson, BHP Billiton Vice-President in Iran, while describing his company as one of a "few visionary companies who believes in the project", is reported to have said (at the Fourth Indian Oil and Gas Conference) that there must be "a shared will to address political issues to allow mutual commercial benefits to prevail." He was of the view that, "the big question is can regional stability be achieved for this to happen." While Samson acknowledged that there were concerns related to security of supply, he proposed mitigating these concerns by implementing two bilateral agreements ensuring continuity of supply – one between Iran and India and another between India and Pakistan. Alternatively, Samson is reported to have suggested one trilateral agreement. According to BHP studies, he confirmed that supplying gas via pipeline – the most affordable long-term energy supply to India – could save India an estimated US$10 billion over 25 years. Iran's massive natural gas reserves could supply India's gas demands for up to 200 years.

The Iranian Option: Geopolitical Feasibility?

Indian Prime Minister Atal Bihari Vajpayee's visit in April 2001 to Iran provided a major boost to Indo-Iranian bilateral ties. On the energy front, Iran has always been a major source of oil for India. The two countries were now looking at the prospects of a long-term energy relationship. During this visit, an agreement for cooperation in the oil and gas sector was signed. Iranian Foreign Minister, Kamal Kharrazi, is reported to have said that laying down the Iran-India gas pipeline across Pakistani territory would contribute greatly to regional peace and stability (Khan, 2001).

The year 2002 witnessed continuing high-level political exchanges between India and Iran, signifying regular consultations on bilateral relations as well as security and strategic issues of mutual concern. Bilateral trade has also shown an encouraging trend of growth. The India-Iran joint working committee on the transfer of Iranian gas to India, constituted in August 2000, has continued with its work of examining various aspects of the proposed India-Iran gas pipeline and other modes of supply of Iranian gas to India. The committee held its fifth meeting at Tehran on 24th and 25th August 2002. The committee reviewed the work relating to feasibility studies, which are in progress at present (MEA Annual Report, 2002-03]. The Indian Minister of External Affairs, Yashwant Sinha, and Foreign Minister of Iran, Dr. Kharazi, met on the sidelines of the United Nations General Assembly in New York on 12th September 2002. The enormous economic opportunities for the export of natural gas to India, and of science-based services and products from India to Iran, seem to provide the imperatives for a 'strategic' relationship between the two countries. New Delhi and Tehran have explored the promising potential for a long–term partnership, especially in sectors such as energy, petrochemicals, fertilizers and information technology and a number of agreements are already in place between the two countries on energy, trade, information technology and customs.

According to Siddiqi (2003: 38), "the cost of building a natural gas pipeline from Iran through Pakistan to India has been estimated at about $ 5 billion. The completion of such a pipeline would have substantial benefits for Pakistan – it has been estimated that it would obtain a total estimated income of $14 billion over 30 years – of which $8 billion would be the transit fee that Iran has reportedly offered the country, plus $1 billion in taxes and $5 billion in energy cost saving." Apparently it looks like a win-win situation for Pakistan and not a zero-sum game, but the debate on pipelines in both India and Pakistan suggests that the proposed pipelines might eventually turn out to be a pipedream in the absence of boundary-producing discourses, and the practices that flow from them, being seriously contested. The persistent instability within Afghanistan, the geopolitical quagmire in Iraq and the designation by President Bush of Iran as one of the "Axis of Evil" countries, further complicate things.

Those who perceive the promise of gas pipelines as some kind of 'peace pipelines' do not hesitate to point to the external push that has accompanied all successful attempts by the two countries to support common projects. We are told that the Indus Water Treaty of 1960 is the most obvious example of the fact that *where there is a will there is a way forward*. The skeptics point out, however, that the case of a joint natural gas pipeline is different in the sense that the pipeline will traverse Pakistan's territory for greater benefit in allegedly 'hostile' India. Therein appears to lie the crux of the problem - namely, the post-partition geographies of mutual suspicion and fear. India feels itself 'vulnerable' to a neighbour bent upon bleeding her through a thousand cuts. The construction of a natural gas pipeline from Iran to Pakistan, if not extended across the border into India, is likely to further strengthen Pakistan's geopolitical orientation towards Western and Central Asia rather than South Asia. It is seen as politically more correct in Pakistan to be seen as collaborating with Iran than with India. In India also many would argue that gas supplies from Iran and the route they should take, must be seen as a bilateral issue, rather than as a multilateral one involving Pakistan.

The prospects of the pipeline projects are closely tied to the extent to which at least three groups in India could be persuaded on this account (Siddiqi, 2003, 52-54). The first consists of the top echelons of the political leadership: especially the Prime Minister, the Petroleum Minister and the Finance and Foreign Ministers. The support of the Opposition Party will also be crucial, given the huge financial and political stakes in the project and the need for bipartisan support, cutting across the ideological spectrum, for such a major undertaking. It is important that this group becomes convinced on the account that commercial interdependencies could create stakes in peace and cooperation, which, in turn, might prove to be the best guarantee against aggressive behaviour on the part of some of the neighbours. Also that the energy security regime, once established, would prove to be sufficiently robust to withstand the ups and downs in bilateral relationships; and that it is in India's enlightened national interest to engage Pakistan and Bangladesh in a cobweb of flows and linkages.

The second group comprises the bureaucrats, including the military bureaucracy, which has traditionally been the hardliner on a vast majority of India-Pakistan issues. With a few exceptions, this group is likely to argue vociferously regarding the availability of an alternate source of energy supply for India and the 'suicidal' consequences of pursuing the Pakistan option. The revival of

Pakistan's good fortune as a 'front line' state for the United States after 9/11, has made this group even more wary of any further 'benefits' that might accrue to Pakistan · from the pipeline, especially the nearly $500 million in projected annual transit fees. It is most unlikely that Luttwak will be able to sell his geoeconomic reasoning to this group, which fears the rise of a pro-pipeline – implying more or less pro-Pakistan – business lobby in India.

The third group is the 'critical' investors group, on which depends, by and large, the prospects for raising funds for the project, and which is expected to establish the liaison with the international consortium for funding, and 'lobby' the political establishment, responsible for both home and external affairs, in the countries concerned for allowing a joint natural gas pipeline with Pakistan. The withdrawal of BHP-Billiton India's Oil Division from New Delhi soon after the Iranian President Khattami's visit to India during January 2003, bodes ill for the role expected from this constituency in realising pipeline projects.

Yet another group that could play a meaningful role in this regard is the mass media, especially when it comes to building bridges of trust by overcoming mental borders. The major question here is how to address the popular fear of intentional supply disruptions of natural gas flows by Pakistan in India. The Pakistani government could interrupt the gas flow at a crucial time, and then blame it on the Jehadis, over whom it anyway claims to have no control. It is also pointed out that, after all, Pakistan does not even abide by WTO regulations and has persistently declined to extend most favoured nation status to India.

As far as India is concerned, a number of forceful arguments have been forthcoming in favour of an on-land gas pipeline reaching India through Pakistani territory. To quote Air Commodore Jasjit Singh (retd), formerly the Director of Institute of Defence and Strategic Analysis, New Delhi:

> Some fallacious arguments are made to say that Pakistan would plough back the $600-odd million earnings from the pipeline into terrorism, as if it does not have the means to do what it has been doing for two decades! But if a long-term view were taken, the rational approach to policy would be to

try and build mutual inter-dependence. Such an approach would start to build stakeholders on both sides to exert pressures for cooper-ative peace rather than search for ways of bleeding through whatever number of cuts. This is also where the international community could play a constructive role.

But the concerns about security of supplies reinforce the political difficulties. Unfortunately, this is one area where emotive intuitive judgments, often without examining the facts, tend to cloud the vision of even eminent people. Security threat to supply routes of oil and gas could occur at the time of war, during prolonged crises and confrontation (like the one last year), and/or interdiction of supplies and sabotage of pipelines by terrorists. Conventional wisdom would have us believe that overland pipelines would be most insecure and vulnerable to such security threats. An objective study would indicate otherwise.

(www.evworld.com/databases/shownews.cfm?pageid=new s230703-05)

Jasjit Singh, as well as a few other analysts, has also argued that an agreement to build a pipeline for the transportation of natural gas must not be seen as a 'bilateral' issue between Pakistan and India. It would require the participation of the producer country (like Turkmenistan or Iran or Oman), plus any other transit country, and international financial institutions (like World Bank and Asian Development Banks who have indicated their interest in funding such projects), as well as Pakistan and, of course, India. Since all of these players would have invested in the project in one way or another, they would have a stake in its continued efficiency. Moreover, Pakistan's own investments would be adversely affected in case of supply disruption. Above all, there are international precedents, potentially binding legal arrangements and safeguards available to guard against supplies being cut off. And it is difficult to accept that only the overland route would be the more vulnerable. In the worst possible scenario, we are told, if Pakistan were to sabotage the gas pipelines, India could play havoc in Pakistan by blocking the flow of the river Indus!

Iran has also been trying to allay fears in India by pointing out that the pipeline would be owned and operated by a consortium, which will be responsible for the supply of gas at the Indian end. In such circumstances, it would be in the interests of Iran and the consortium to ensure that Pakistan abides by the agreement and stipulate penalties for violations. If Islamabad agrees to this, and its public pronouncements suggest it will, then it should logically also examine the irrationality of continuing to deny to India MFN status. As for India, there is no evident harm to exploring the proposal critically, provided measures necessary to reduce the risks are incorporated, and insurance taken out against possible interruptions of supply. Gas and oil pipelines even between countries which are adversaries have worked elsewhere. For example, during the height of the Cold War, the Soviets extended their pipeline to Western Europe against opposition from the USA. The argument then becomes that a breakthrough in India-Pakistan relations on the issue of pipelines would also have a positive impact on Bangladesh, which has been resisting selling its gas to India next door. Rather than reject the pipeline outright on the grounds that it traverses hostile territory, India should adopt a more nuanced attitude towards the project.

A few among those questioning the political wisdom behind letting the on-land pipelines pass through the soil of a hostile neighbour have expressed their preference for an undersea option. This option, they believe, will completely deny any leverage whatsoever to Pakistan. It seems that they are missing a few points. First and foremost, as pointed out by Jasjit Singh:

> . . . an undersea pipeline (whether in deep sea or in shallow 50 metres depth close to the Pakistani coastline) could be blown up by an enemy with few, if any, safeguards available to India; the time for repairs and their costs would be highest for the undersea route and the least for overland pipelines. Disruption in case of war would affect all options. Offshore routes would actually give a free hand to the ISI to wreak havoc on our economy. Contrary to popular belief, liquefying natural gas, transporting it on special tankers, de-liquefying it on our shores and then pumping it across the land mass would not only be costly, but would remain vulnerable. Pakistan's geography allows it adequate opportunity to

interdict such traffic easily from bases on the Makran Coast. And it has built up one of the largest maritime strike capabilities in the Indian Ocean. The problem during war and prolonged crisis would be acute as the tanker war in the Persian Gulf during 1986-88 showed. Terrorist attacks like the one on USS Cole with small boats would pose additional risks without the safeguards available in the case of overland pipelines.

Compared to that, Pakistan would have to provide access to regular surveillance and monitoring of the overland pipeline by a multinational agency and immediate access to assess and repair any damage. There would be residual risks; but damage to gas pipelines overland are normally repaired within 2-7 days. Discoveries of gas reserves on the eastern seaboard would help reduce the dependency on the western routes reducing somewhat their value as targets for interdiction. (Ibid.)

Yet another aspect that demands and deserves attention while talking about submarine pipelines relates to maritime jurisdiction. Should India opt for a pipeline that hugs the coastline but passes through the EEZ of Pakistan or, making full use of the technologies available, go for a pipeline located outside the EEZ? There have also been objections from certain quarters in the Pakistan Navy that such a pipeline will allow the Indian Navy to get too close.

A number of approaches were identified by the group mentioned earlier in this chapter, which had met in Udaipur, India, in 1996 to examine the steps that could be taken to ensure that there were no intentional attempts to sabotage gas flows. These deserve the serious attention of skeptics as well as those outrightly opposed to the whole idea of the pipeline from Iran to India through Pakistan. Reproduced below from Siddiqi's study (2003: 57-59) are several approaches identified by the group:

1. The amount required to build a pipeline from Iran to India through Pakistan is about $5 billion. Such a large amount would require international financing, which could be guaranteed by multilateral organizations such as the World Bank or the Asian Development Bank or by commercial

banks with guarantees from their own governments. These institutions could require building guarantees from Pakistan that there would be no intentional disruption of supplies to India.

2. The risk of sabotage of the pipeline by individuals and groups pursuing their own goals exists for both Pakistan and India. It can be reduced by taking actions that all natural gas transmission companies take, including patrolling and remote monitoring, but the risk can not be eliminated. Such disruptions are usually of relatively short duration. The important factor here would be that the Pakistani government did not encourage the sabotage, and that the latter was also being adversely affected. A working group consisting of representatives of users of gas in both countries and of the operators of the pipeline could be set up to design measures to enhance security.

3. A multi-party agreement involving Iran, Pakistan, India, and possibly other countries financing the pipeline could be signed that curtailment of the natural gas deliveries to India would result in curtailment of deliveries to Pakistan as well.

4. A 'Take or Pay' clause could be incorporated into the purchase agreement stating that Pakistan would pay for the entire amount of gas supplied, even if it were not transmitting it further to India. Since about 2/3 of the total gas would be going to India, it would be rapidly beyond Pakistan's financial resources to continue to pay for this additional gas.

5. Some power plants using the natural gas could be located in India close to the Pakistan border, and would supply electricity to parts of both countries. Any disruption of supplies to these power plants would adversely affect both countries.

The above-mentioned steps to 'securitise' gas flows are no doubt thought provoking and at least deserve serious attention of the intellectuals and institutions of statecraft in the countries concerned. But it seems they have overlooked certain categories of fault lines related to Pakistan's domestic politics, not directly related to the question of pipelines.

Fault Lines in Pakistan

Let me now turn to fault lines within Pakistan. Despite Pakistan's cabinet approval of an on-land pipeline to India, there appears to be an in-built resistance in the political system, reinforced by a dominant governmentality, which is obsessed with the idea of seeking parity with India in all walks of life and looks at the problem of Jammu & Kashmir as an unfinished task of partition. It is understandable, therefore, if some amongst the military establishment argue that any concessions given to India on the energy security front are likely to undermine Pakistan's stand and support on the so-called 'core' issue of the Kashmir dispute. Yet it is reasonable to assume that since the project has received the approval by a military regime in the cabinet, the senior-most body involving military and civilian participants, the concerns of the recalcitrant sections in the army have been overcome to a large extent. All said and done, however, the crucial task of ensuring the physical security of pipelines cannot be accomplished without the goodwill and positive attitude on the part of the military and paramilitary forces.

Having said that, it is also important to bear in mind that the space of flows is not devoid of socio-cultural content. One example where culture may experience transformation by way of economic development is in the actual construction of the pipeline and the route it follows. These two issues deal directly with the environment and people of indigenous communities. Travelling from Iran and into Pakistan, the pipeline project would encounter various tribal communities in the rural areas of Baluchistan, where a large part of the pipeline will be constructed. From a socio-cultural perspective, a wide variety of conflicts may emerge from this encounter.

First, the construction of such a lengthy pipeline will employ large amounts of local labour. While this will provide temporary employment for many of Pakistan's rural poor, it may also interfere with domestic labour dynamics. For example, the Province of Baluchistan has experienced a constant influx of Afghan refugees for the past 20 years due to Afghanistan's ongoing civil war. These refugees, poor and illiterate, are top choice over Pakistani labourers for local businesses as cheap labour. They will work for lower wages than Pakistani labourers. As a result, increased tensions exist between the Afghan labourer who is seeking refuge and stability in neighbouring Pakistan and the Baluch labourer who is trying to feed his/her family while living on a meagre wage. Both groups seek to benefit from potential pipeline construction in their regions. However, the question we need to ask is which group will be privileged enough to be seen as the cheaper form of labour?

While the physical construction of the pipeline will require large amounts of unskilled labour, other technical aspects of pipeline implementation will require skilled and educated workers with backgrounds in science and technology. These workers may be recruited from both abroad and within the country of construction. Due to the increased focus on careers and education in science and technology in recent decades in both India and Pakistan, employment of skilled workers will bring much-needed jobs to an educated middle class. Second, since the pipeline will travel through rural areas in Pakistan, it will encounter areas of land controlled by feudal landlords or tribes. The Pakistani government may choose to pay off these groups by unofficially granting them access to the natural gas resources and/or some form of control over the pipeline. Another option is that the government may officially purchase the land from these groups. Either way, such options can alter or come into conflict with the socio-cultural mindsets of tribal communities. For example, a certain tribe may have an historical, political, or familial tie to the land. Also, the distribution of more land to one family may grant them more political power than a family that has less land. The selling of land by certain groups may disrupt the political structures of tribal communities.

Lastly, the development, construction, and continued use of the pipeline may transform the lifestyles of indigenous rural communities in both Pakistan and India. In Pakistan, there is already discontent over the proposed cultivation of natural gas reserves in the Kirthar Wildlife National Park in the southern Province of Sindh. Alongside international environmental groups, the Sindh Wildlife Department and indigenous people living in the area protested over gas exploration in the park. Gas exploration will not only damage the natural environment of the park but may impact on the lifestyles of indigenous peoples. The cultivation of natural gas reserves could transform and displace the agricultural economy of rural communities, replacing it with a situation where cheap labour is required for the construction of the pipeline and other industrial purposes. The call for cheap labour will inevitably result in the migration of workers from other rural communities to places where pipeline construction and industrialization are occurring. This, in effect, will result in dramatic shifts in the local economies of rural communities. They may no longer be able to depend on agricultural production and labour if there is a shift in the employment prospects in nearby locales. Construction and industrialisation are not negative forces in the context of development, especially in poor countries. It is important, however, for governments and companies to accurately and fairly assess the cost of such activity on indigenous rural communities.

The issues mentioned above no doubt will have to be addressed by both state and non-state agencies involved in the joint business of ensuring energy supplies. One critical issue-area that demands and deserves immediate attention as well is the geopolitical turmoil that manifested itself dramatically on 11 January 2005 in the Baluchistan province of southern Pakistan; an extremely poor Province through the territory of which the proposed Iran-India gas pipeline will traverse for about 750 kilometers. On that fateful day, the Bugti tribesmen and the elusive Baluchistan Liberation Front (BLF) attacked Pakistan's largest gas-producing plant in Sui that accounts for nearly 45 per cent of total national gas production. In recent years, this part of the Baluchistan-Punjab border has been a battleground of private militia belonging to Baluch tribes, targeting water pipelines, power transmission lines and gas installations.

These geopolitical rivalries, described early on in this chapter, run rather deep in a region that is strategically important due to its large reserves of oil and gas. They are also the outcome of a longstanding popular perception among the Baluch tribes that they have been persistently denied a fair share of the oil and gas wealth by the federal government. The net result of such power asymmetries and regional imbalances within the Pakistan polity has often been violence and counter-violence from time to time, on the one hand, and a persistent resistance by the Baluchi people to 'developmental' projects. One might recall that it was on the night of 8 January 2003 that the Baluch Liberation Front (BLF) fired rockets at the pipelines disrupting supply to a power plant. In a separate incident, a pipeline close to Sui township, 250 miles north of Karachi, was attacked. The recurrence of developments such as these in January 2005 has cast some serious doubts in India and elsewhere over the feasibility and reliability of the project.

Bangladesh-India Gas Pipeline: Exporting Sovereignty to the Other?

India has also been looking eastwards for its supply of natural gas. Both Bangladesh and Myanmar have significant reserves of natural gas and have been identified as potential suppliers. In fact, India and Myanmar have agreed to exchange geo-technology data of basins lying across the India-Myanmar border so that both sides can optimize their exploration programmes in this basin, and ONGS has also evinced interest in studying exploration opportunities in Myanmar (Dadwal, 2002). More recently, India has secured a 30 per cent equity participation in the exploration of gas off the Myanmar coast. Among several other options available to transport this gas to India, the one that is now being seriously considered by New Delhi is the pipeline of around 290 kilometres through Bangladesh territory. However, "Bangladesh has demanded that India should give transit rights to its territory from Bhutan and Nepal, facilitate electricity supplies from these countries and also grant unilateral trade concessions to enable it to balance its trade deficit with India, before its considers the Indian proposal for a pipeline through its territory" (Parthasarathy, 2005). Moreover, Minister of State of Bangladesh for energy, A.K.M. Mosharraf Hossain, proclaimed: "If India does not allow these, we shall not sign any tripartite agreement (on a gas pipeline)" (Ibid.). One image of Bangladesh, even though highly contested within the country, is that it is "floating on a sea of gas". The debate on the feasibility and desirability of energy exports to India revolves around the question of how 'abundant' or 'scarce' this strategic resource is or is going to be in the decades to follow. Ultra-nationalists in Bangladesh spare no opportunity to look and sound worried over the fact that there is simply not enough for future domestic consumption. Pragmatists, far lesser in number and influence, uphold that there is enough for domestic use and some extra for export. Realists, numbering a hopeless minority and bordering extinction, remain painfully aware that that unless

Bangladesh exports some of its gas, it will not be able to able to pay the oil companies in dollars. According to official figures released by Dhaka, gas reserves are around 16 trillion cubic feet. The US Geological Survey in a recent report has estimated that there is an additional 32 tcf of 'undiscovered reserves'. Whereas the oil companies have been most vocal while flagging the point that without a political will and decision to export gas to India, prospects for new exploration, which could help expand the size of 'known reserves', remain minimal. In short, what we are actually witnessing in Bangladesh is a highly contested site marked by numerous ideological and geopolitical fault lines.

The country has had 22 natural gas and one oil well discovery over a century-long exploration activity, with 68 discovery wells being been drilled so far. Its success rate is said to have been around 34 per cent. The Bibiyana gas field is believed to be one of the richest and most promising in the world. Though estimates of natural gas reserves in the country vary from the moderate to the unrealistic, as noted above, it is generally believed that Bangladesh has sufficient gas to not only satisfy its domestic requirements but have enough left over to export a substantial amount, thereby bringing in the urgently needed foreign exchange. According to the latest estimates reported by the Norwegian Petroleum Directorate for the sate-owned *Petrobagla*, potential gas reserves are around 64 tcf, giving Bangladesh about 160 years of gas supplies.

The fact that the energy security debate in Bangladesh has also fallen into some kind of a territorial trap was revealed through a dialogue organised by the Centre for Policy Dialogue, Dhaka, on the theme, "Restructuring the Energy Sector: Current Issues", on 29 May 2000. The dialogue was attended by a cross-section of politicians, bureaucrats, professionals, academicians, business corporations and representatives of the media and civil society. The participants were to emphasise at the very outset, "the need for independent regulation and monitoring in order to conduct realistic and authentic analysis of energy sector and formulate effective and transparent policies." The stand of the government on the whole issue of energy exports was being described by the Minister for Industries, Tofail Ahmed, as 'negative', at least from the short-term perspective. As a principle, he affirmed, "the government would like to be very cautious in this matter in order to avoid a fate like Nigeria, which had been unable to reap benefit from its vast natural resources". On the other hand, there were those who argued that the example of Nigeria was not relevant, since the governance in Bangladesh is much better. It was further pointed out that the debate on energy exports should be conducted on a *more rational and economic note*, without anybody breaking out in an avalanche of political rhetoric.

It was pointed out by one of the delegates that a basic problem is that gas as an export commodity is a new concept to the general public. The fear in the public mind, it was argued, was two-fold: one is that, since it is an exhaustible resource, we cannot regenerate it once we have exported. The other problem is the lack of public confidence in the financial and political system, which makes the possibility of a huge amount to be generated in the system a frightening notion for the public, who fear gross mismanagement of the accrued funds.

It is to state the obvious perhaps that the geographies of 'fears' can at times act as fault lines, that are not easy to bridge. They can and do turn geographical proximities into geopolitical divides and distances. This is precisely the point that many analysts tend to miss while stating that it is so 'obvious' that, for Bangladesh, India is a 'natural' market; therefore, by building a pipeline from Bangladesh, and linking it with the HBJ pipeline, much-needed gas could be brought to the energy-starved states of Haryana, UP, Bihar Punjab and Rajasthan as well as Delhi. Whereas the fact of the matter is that, for the past several years, New Delhi has been trying hard to persuade Dhaka that it should be selling gas to India, even offering to locate, explore, tap and enter into a purely commercial venture. However, Dhaka continues to prevaricate on the grounds that it has to first ensure that it has enough reserves to serve the domestic market for the next 50 years before it can export. While Bangladeshi politicians are still debating whether to calculate their gas reserves on the basis of 'proven', 'probable' or 'possible' reserves, they are, consciously or unconsciously, testing the patience of the foreign operators – already running in short supply it seems – who are running up debts due to delays. According to the hardliners in Bangladesh, they would not like to 'export their sovereignty' to

India even if the country were to lose a huge ready-made market next door.

A major non-state actor on the energy scene of Bangladesh has been Unocal. Unocal is the same company that wanted to build a natural gas pipeline from Turkmenistan to India through Afghanistan and Pakistan. They tried hard to sell the idea to the Taliban, which was more interested in spreading the Islamic revolution than in making money. Amidst protests that it is dealing with the devil, Unocal finally pulled out of the project in the late 1990s. Unocal continues to believe that eventually a network of pipelines will physically connect the producers and consumers of oil and natural gas in Asia from Indo-China to Turkey. It has worked in Bangladesh since 1997, in Bibiyana, and increased its operation following its withdrawal from the pipeline project in Afghanistan. It has invested millions of dollars in the project, and has even invested small amounts in every hospital, road, school and telephone facility that Bangladesh has set up in the last few years; now it is anxious to get some return on its investment. In 1999, significant deposits were struck and Unocal has been pressuring Dhaka to allow it to export some gas, mainly to the huge and eager Indian market.

Unocal wants to build a 1,363-km pipeline to move gas from the Bibiyanah gas field in North Eastern Bangladesh to the HBJ pipeline near Delhi. It will run through West Bengal, Bihar, and Uttar Pradesh. Unocal has a simple case: the $1.2 billion project would bring in an estimated revenue of $3.7 billion to Dhaka over the next 20 years, and an immediate investment of about $700 million. Jobs and development associated with such a mega project are said to be the bonus. Unocal also believes that Bangladesh could become an energy hub allowing the transport of gas from Myanmar and North Eastern India to the energy-hungry mainland. Dhaka could just sit back and charge transit fees.

More recently, the "National Committee to Protect Oil-Gas and Power-Port" is said to have blamed US Ambassador, Harry K. Thomas, for violating diplomatic norms in talking about gas exports to India (*The New Nation*, Internet Edition, 9 October 2003). Announcing a fresh programme of long march for October 19-25 towards Mongla, the committee also alleged that the envoy launched an 'evil effort for the pipeline of gas'. Without sourcing, it cited press reports alleging that US Embassy officials lobbied for gas exports and a private port when they attended some social parties in Chittagong. The Committee leaders, addressing a press conference, are reported to have vehemently criticised, "Shell-Cairn's bid to conduct an aeromagnetic survey in Sunderbans, to destroy its ecosystem." It was further alleged that, "The Energy State Minister has given permission for the survey although the Environment and Forest Minister announced that it wouldn't be carried out", said a written statement of the committee.

Religious Fault Lines: Clash or Dialogue of Civilisations

Last but not the least, it appears that the discourse as well as practice of energy flows in South Asia will have to encounter the juxtaposition between the 'clash' and 'dialogue' of civilizations against the backdrop of rising religious 'fundamentalism' in various parts of South Asia. How is geoeconomics going to encounter and negotiate with the perception, likely to be entertained by certain sections of the Islamist movements in Iran and Pakistan, that the involvement of Western companies in the development and transportation of energy resources is an extension of Western exploitation of the East or yet another neo-imperialistic ploy. Also, various political-religious movements may associate Western involvement with ideas of allegedly 'anti-Islam' secularism. The oil and gas pipelines may now be imbued with news meanings and symbols and perceived as a threat to the project of implementing national infrastructures for the application of interpretations of Islamic law, values and traditions. Accordingly, the Islamists may not view economic and natural resource development by Western interests as the most viable solution to economic deterioration, lack of sustainable development, and the looming energy crisis in South Asia.

How does one interpret the willingness – bordering compulsion – on the part of various of the Western oil firms competing with each other for contracts to build oil and natural gas pipelines from Central Asia through Afghanistan and into South Asia by courting – openly as well as secretly – the governing Islamist Taliban regime in Afghanistan? Apparently, here is an encounter between the two divergent forces and value systems. The mere fact that these two entirely different world views, are interacting with one another is evidence enough of the intersection of trade and culture and its impact on the interaction between groups around the world. This example has the potential to duplicate itself in Pakistan as well, where Taliban-influenced philosophies exist and where anti-imperialist movements thrive in the form of Islamist political parties. What kind of tone and toner this dialogue is going to acquire is difficult to say at present.

Conclusion

South Asia, held hostage for many decades now to India-Pakistan conflict, especially over Jammu & Kashmir, poses formidable, though not entirely insurmountable, challenges for new collaborative projects. This eco-geographical region would indeed do well to become a 'space of flows' where joint projects focused on development, and cooperation could help in transforming closed and hostile borders into trans-border zones of cooperation. A natural gas pipeline agreement between India and Pakistan would be a major confidence-building measure between India and Pakistan, and could lead to reduce tensions in other fields, as well as facilitate other projects in the sub-continent such as the Bangladesh-India pipeline.

Ideally speaking, if the geoeconomic logic of commercial interdependencies were to dictate and drive the destiny of Southern Asia(ns), both energy-supplying and energy-consuming regions, through their respective governments, could jointly negotiate frameworks within which commercial enterprises could settle terms and conditions pertaining to quality, quantity, price, and other factors. The government concerned could also agree on measures relating to non-commercial risk such as natural disasters, political tension and terrorism. There could be an agreement between government and the parties concerned on investment and protection, currency transfer, loan security, and other issues, which would create a suitable legal regime for the protection of commercial interests. A joint mechanism could be established by the countries concerned to supervise project implementation and operational matters including construction issues, production facilities, supply sharing, transit arrangements, and environmental matters. Various agreements on modalities for the settlement of disputes related to these projects could also be put in place. The terms and conditions for the tender/bid process and the award of contracts, including prequalification and eligibility criteria could be mutually agreed through dialogue and negotiations. A time frame and period of validity for the inter-governmental agreements covering the commercial contracts could also be established. In short, this is how an energy security complex for Southern Asia could look like, provided the intellectuals and institutions of statecraft in South Asia were to look for and find new geoeconomic roles.

How should we treat the flows? Do they pose opportunities or threats? Or both? And to whom? Do the pipelines unite or divide peoples and places? It appears that they do both, at least in South Asia. Much depends, as illustrated in this chapter, on the larger and deeper ideological, political and socio-cultural contexts in which the pipelines are embedded and implicated. Here are some of the opportunities, for example, for both India and Pakistan. If Pakistan, like India, opts for natural gas as a new source of renewable energy for its domestic market, then it may be able to resolve the conditions of lack of access to rural communities. The pipeline would be travelling from Iran from the southwestern portion of Pakistan towards Multan, an urban settlement located in the heart of Punjab Province. The land between the southwestern Pakistani-Iranian border and Multan is predominantly a desert and a dry area populated by tribal communities living in villages. It is proposed that the pipeline will be opened for domestic use in Multan. However, the fact that it travels through remote rural areas where renewable energy is in demand means that prospects exist for extending the pipeline into a domestic network providing natural gas to village populations. Thus, the proposed pipeline has the potential to promote renewable resource development/deployment and improve energy efficiency.

No doubt, the fault lines – both material and symbolic – persist in 'South Asia' at various levels, and in some cases have acquired an extraordinary complexity and rigidity. However, at the same time, it is difficult to deny the growing appeal as well as desirability of transforming South Asia into a geoeconomic 'space of flows', both among state and non-state actors. It is becoming increasingly doubtful whether territorial borders in South Asia will be able to fixate and regulate mobility and flows associated with the inescapable imperatives of energy-ecological security.

I have also tried to argue and illustrate in this Chapter that the notion of 'fault lines' needs to be conceptualised in a dynamic, rather than static, sense of Othering and (B)ordering. The experience of post-colonial, post-partition, South Asia strongly suggests that bordering processes neither begin nor end at demarcation lines in space or time; rather, they symbolize a social practice of spatial differentiation. The discourse and practices of energy (in)security in South Asia are deeply implicated in the territorial strategies of ordering, bordering and othering, which often take place, although certainly not necessarily, at the spatial scale of states. I have shown elsewhere (Chaturvedi, 2002) how the discourses and strategies practiced between India and Pakistan, where practices of inclusion and exclusion are framed by the nation-building projects of two countries, are discursively uttered through differences in religion. How, for example, education programmes in Pakistan reproduce and reinforce otherness.

This Chapter has also shown that geopolitical concepts such as 'fault lines' do not just reflect the boundary-producing practices that took place in the past but also constitute the current political discourses through which new region-building processes and initiatives are invented and/or thwarted. Not all the fault lines in South Asia, however, are by default! Unfortunately, while constituencies for peace-building are scarce, the number of stakeholders in perpetuating conflicts – especially in the partitioned Jammu & Kashmir – has increased ever since the 'bloody' partition of British India. Making others through the territorial fixing of order, has unfortunately become intrinsically connected to the dominant image of borders in South Asia. 'Others' have become not only necessary but also useful, and therefore constantly produced and reproduced to maintain 'cohesion' and 'integrity' in territorially demarcated societies. In several respects, South Asia continues to be imprisoned by what Ohmae has termed the old cartography, a cartography that cannot map the globe-girdling network of flows in the form of corporations, trade and communications infrastructure, and oil and gas pipelines. The challenge before India and the rest of (south) Asia is to break through this prison of old cartography and territoriality.

References

Alam, S. 2001. "Iran's Hydrocarbon Profile: Production, Trade and Trend." *Strategic Analysis* 25(1) April: 119-133.

Ali, M.J. 1983. "The Behaviour of the Oil Market: Cycles and Structures." *OPEC Bulletin* 14(10) November: 17-19.

Bakshi, G.D. 1997. "The Energy Security Scenario: Focus on the Middle East." *Strategic Analysis* 20(1) April: 59-69.

Bansal, R. 2002. "MEA Offers Economic Diplomacy Input from Missions Worldwide", *The Financial Express*, New Delhi, 15 February.

"Bangladesh Could Have Huge Natural Gas Reserves". *The Times Of India* 16 February 2000: New Delhi.

Bhatia, R. 1982. "Surviving the Oil Crisis, the Case of Oil Importing Developing countries." *Yojana* 26 January: 64-65.

Chaudhary, R. R. 1998. "Energy Security Policy for India: The Case of Oil and Natural Gas." *Strategic Analysis* 21(11) February: 1677.

Chaudhary, R. R. 1998. "An Energy Security Policy for India: The Case of Oil and Natural Gas." *Strategic Analysis* 21(11) February: 1671-1682.

Chaudhury, R. R. 1995. "Prospects for International Pipeline in the Indian Ocean." *Journal of Indian Ocean Studies* 2(2) March: 132-136.

Chaudhury, R. R. 1993. "Indo-Omani Gas Pipeline: Major Security Issues at Stake. *Indian Express* 4August: New Delhi.

Chaudhary, S. 2000. "Iran to India Natural Gas Pipeline: Implications for conflict Resolution and Regionalism in India, Iran and Pakistan." *Trade and Environment Database Case Studies*, American University. Washington D.C.

Cherian, J. 2001. "Mixed Signals in Tehran." *Frontline* 18(9) April-May: 50.

Dadwal, S.R. 1998. "Politics of Oil: Caspian Imbroglio." *Strategic Analysis* 22(5) August: 751-760.

Dadwal, S.R. 2002. *Rethinking Energy Security in India.* New Delhi: Knowledge World.

Dalby, S. 1992. "Security, Modernity, Ecology: The Dilemmas of Post-Cold War Security Discourses". *Alternatives: Social Transforma-tion and Humane Governance* 17(1): 95-134.

Datta, S. 2002. "Indo-Bangladesh Relations: an Overview of Limitations and Constraints ." *Strategic Analysis* 26(3) July-September: 427-437.

Diwanji, A.K. 2000. "Geo-Political Issues set to Dominate Proposed Gas Pipeline from Iran to India". *The Rediff Business Special*, 13 April.

Edward, L. M. 1999. " The New Political Economy of Energy In Fueling the 21st Century" *Journal of International Affairs* 53 (1) Fall.

Ferruto, E.M. 1983. "The Oil Market, Conjuctures and Outlook." *OPEC Bulletin* 14(10) November: 19-20.

Ferrel, J.C. 1977. "Strategy for Alternative Energy Sources." *Journal of Alternative Sources, an International Compedium* 10 May: 4829-4833.

Fesharaki, F. 1999. "Energy and The Asian Security Nexus". *Journal of International* Affairs 53 (1) Fall: 79-80.

Gas Authority of India, 2000. *Limited 15th Annual Report* 1998-99. New Delhi: GAIL.

Ghorban, N. 1996. " Middle East Natural Gas pipeline Projects: Myth and Reality".*The Iranian Journal of International Affairs* Fall: 649.

Gobble, P. A. (1995), "Pipelines and Geopolitics, *Prism*, 1(23). F:\Paper on Pipelines\Geopolitics of Pipelines\Pipelines and Geopolitics Prism, November 3, 1995 _Jamest.htm.

Government of India, 2000-01. *Economic Survey.*

Harakavy, R. 2001. "Strategic Geography and the Greater Middle East." *Naval War College Review* 54(4): 37-53.

Hodgson, P. E. 1999. "The Renewable Energy Sources" *In his Nuclear Power, Energy and Environment*. London: Imperial College Press.

Hodgson, P.E. 1999. *Nuclear Power, Energy and Environment*. London: Imperial College Press.

IEA, 1998. *International Energy Outlook.*

Imai, R. 1997. "Energy Issues in Asia in the Twenty First Century". *Asia-Pacific Review* 4(2) Fall/Winter: 227-228.

"In the Pipeline: Indo-Bangladesh Power Play", 2001. *Business Standard*, 26th October.

"India Can Join Gas Pipeline Project". 1995. *The Times of India* 20 April: New Delhi.

Johar, H. and Bhagat, G. 1995. "The American Dilemma in the Gulf Region". *Journal of South Asia and Middle East Studies* 19(1) Fall: 277-278.

John, G. G.2000. "Coal and Energy for the Twenty-First Century in India", in Audinet, P. Shukla, P.R. and Gare, F. (eds.), *India's Energy Security: Essays on Sustainable Development*: New Delhi.

Kemp, G. 1998-99. "The Persian Gulf Remains the Strategic Prize". *Survival* 40(4) Winter: 39-40.

Khan, M.N. 2001. "Vajpayee's Visit to Iran: Indo-Iranian Relations and Prospects of Bilateral Cooperation." *Strategic Analysis* 25(6) September: 765-777.

Khouja, M.W. (ed.), 1981.*The Challenges of Energy, Policies in the Making*. New York: Longman.

Kiani, K.2001. "Slim Prospects of Trans-Pakistan Gas Pipeline" *The Dawn* 12 April.

Klare, M. 2003. "New Geopolitics", *Monthly Review*, 55(33).

Kumar, M. "India Gears Up for Energy Security: Regional Cooperation in the Offing", *The Tribune*, Chandigarh, 29 January.

Lacoste, Y. 2000. "Rivalries for Territory." *Geopolitics* 5(2): 122-158.

Lahererre, J. 2000. "Oil and Gas Production Compared: Past and Future" *Tomorrow's Oil* 2 (10) October: 22-27.

Luttwak, E. N. 1990. "From Geopolitics to Geo-economics, Logic of Conflict, Grammar of Commerce." *The National Interest* 20: 17-24.

Mamadouth, V. D. 1998. "Geopolitics in the Ninties: One Flag Many Meanings." *Geo Journal* 46: 237-253.

Mann, P. 2001. "Fighting Terrorism: India and Central Asia.' *Strategic Analysis* 24(11) February: 2047.

Mohan, C.R. 1996. "Geopolitics and Energy Security." *Strategic Analysis* 19 (9) December: 1269-1276.

Mohan, R. 1996. "Central Asia's Pipeline Dreams". *The Hindu* 12 October: Delhi.

Morse, E. L. 1999. "A New Political Economy of Oil?" *Journal of International Affairs* 53(1) Fall: 24.

Mukherjee, S. 2001. "Unocal plans Natural Gas line from Bangladesh to Delhi". *Economic Times* January: New Delhi.

Naaz, F. 2001. "Indo-Iranian Relations: Vital Factors in the 1990s", *Strategic Analysis* 25(2) May: 227-240.

"Oil Pipeline Afire after 'Sabotage'." 2003. *The Hindu* 14 August: New Delhi.

Pachauri, R. K. 1985. *The Political Economy of Global Energy*. London: John Hopkins University Press.

Pachauri, R. K. 2000. "Not a Pipe dream: Back to Iran Gas Pipeline Option," *The Times of India*, 17 August: Delhi.

Parekh, J. K. 1980. *Energy System and Development*. Delhi: Oxford University Press.

Parikh, J.K. 1985. "Modelling Energy and Agriculture Interaction." *Journal of Energy* 10(7) May: 793-804.

Parthasarthy, G. 2001. "Development in Indian Ocean Region: New Delhi Quest for Energy Security". *The Tribune* 5 July: Chandigarh.

Parthasarthy, G. 2005. "Pipelines or Pipe dreams? Security Considerations Ignored", *The Tribune* (Chandigarh) 24 February.

Parthasarthy, G. 2001 "Some Policy Options for India", in Singh, J. (ed.), *Oil and Gas in India's Security*. New Delhi: Knowledge World

"Polls Menace Bangladeshi Gas Export Hopes,' 2002. *Petroleum Intelligence Weekly*, 25 February.

Parthasarthy, G. 2001. "*Some policy Options for India*" In Singh, J. (ed.), *Oil and Gas in India's Security*, New Delhi: Knowledge World: 107-115.

"Qatar to Pakistan Gas Pipeline: MOU Singed". Report on 19 July 2000.

Rao, A. R. 1985. "Energy consumption in Rural Transportation in India." *Journal of Energy* 10 July: 681-682.

Razavi, H. 1998. "Investment and Finance in the Energy Sectors of Developing Countries." *The Emirates Occasional Papers* 22 June: 17-18.

Reedy, B.M. 2000. "India-Iran Gas Pipeline: Pakistan Seeks to Allay Indian Security Concern" *The Hindu* 8 September: Delhi.

Richardson, B. 1999. "Geopolitics of Energy in to the 21st century," *Strategic Energy Initiatives*, CSIS Washington: 8–9 March.

Rizvi, S.A. 2000. "$3 Billion Project of Iran, Pakistan and India Gas Pipeline", *Pakistan and Gulf Economist* 25-31 December: 39.

Russell, J. 1983. *Geopolitics of Natural Gas*. Massachusetts: Ballinger Publishing Company.

Sachdev, R.K. and Maggo, J.N. 1996. "India Energy Overview", Paper presented at the seminar on *Asia Energy Vision 2020* in New Delhi.

Salameh,M.G. 2003. "The New Frontiers for the United States Energy Security in the 21st Century", *Applied Energy* 76: 135-144.

Salameh, M.G. 2003. "Quest for Middle East Oil: The US versus the Asia-Pacific region", *Energy Policy* 31: 1085-1081.

Sharma S. 1983. *Development Strategy and the Developing Countries*. New Delhi: South Asian Publishers Ltd.

Siddiqi, T.A. 1995. "India-Pakistan cooperation on Energy and Environment: To Enhance Security". *Asian Survey* 35: 280-290.

Siddiqi, T. A. 2003. *Enhancing Clear Energy Supply for Development, A Natural Gas Pipeline for India and Pakistan*. BAUSA for Peace.

Siever, R. 1980. *Energy and Environment*. San Francisco: Freeman and Company.

Singh, J. 2001. *Oil and Gas in India's Security*. New Delhi: Knowledge World.

Singh, K.R. 2002. "Geo-strategy of Commercial Energy". *International Studies* 39(3): 259-288.

Socor, V. 2002. "A Pipeline Pact that could keep the Peace", *The Wall Street Journal*, reprinted in *Indian Express* 8 June: Delhi.

Song, J. 1999. *"Energy Security in the Asia-Pacific: Competition or cooperation."* Report on the Seminar organized by the Asia-Pacific Center for Security Studies at Honolulu, Hawai on 15 January.

Sparke, M. 1998. "From Geopolitics to Geoeconomics: Transnational State Effects in Borderlands", *Geopolitics* 3(2): 62-98.

Supply of Hydrocarbon Vision 2025, http// www.pinindia. nic.in/pinintiative/ hydro_rv2d.htm.

Tata Energy Research Institute 1998. *TERI Energy Data Directory and Yearbook 1998/99*. New Delhi: TERI.

Tata Energy Research Institute 2000. *Overview of the Indian Energy Sector*. New Delhi: TERI.

TERI Energy Directory Yearbook. 1996-97:71.

Tongia, R. and Arunachalam, V.S. 1999. "Natural Gas Import by South Asia", *Economic and Political Weekly*: May 1.

Verghese, B. G. 2001. *Reorienting India: The New Geopolitics of Asia*. New Delhi: Konark.

Young, O.R. 1989. *International Cooperation: Building Regimes for Natural Resources and the Environment*. Ithaca and London: Cornell University Press.

CHAPTER 8

The Geopolitics of Japan's Energy Security

Dennis Rumley

Introduction

Unlike other Northern states of the "Industrial Triad", Japan has few energy resources and the development of a 'hemispheric' strategy along the lines of the United States is also highly problematical due to regional competition for energy resources from other northeast Asian states, especially China. As a result of these pressures, Japan is therefore inevitably pushed towards a greater dependence on untapped energy reserves in the developing world around the Indian Ocean and elsewhere, a greater reliance on nuclear energy and a much greater investment into alternative energy sources, such as hydrogen and nuclear fusion. Japan's 'necessity' in this regard may well mean that, in the long-term, it becomes energy self-sufficient as well as a world leader in energy technology.

It is assumed here that access to long-term stable supplies of energy is one of the principal driving forces in the foreign policies of many states, especially those of the industrial triad. In support of this assumption, this chapter will first briefly outline some of the main characteristics of Japan's energy insecurity. Second, the chapter will describe the structure of Japan's Indian Ocean energy dependence. Third, it will discuss some policy ideas to develop an

East Asian energy strategy. Fourth, the chapter will consider some of the issues which link energy security and environmental security in Japan. Finally, the chapter will discuss some of the more relevant policy concerns for Japan's energy security future.

Japan's Energy Insecurity

In 2002, Japan was the world's third largest consumer of oil after the United States and China; it was the sixth largest consumer of natural gas; it was the fourth largest consumer of coal and it was the third largest consumer of nuclear power (BP, 2003). Although its per capita consumption of electricity is less than many states, including Australia, Canada, Finland and Iceland, nevertheless it is the world's fourth largest overall consumer of energy (5.4%) after the USA (24.4%), China (10.6%) and Russia (6.8%). The essential energy problem for Japan, compared with these other very large consumers is that, at present, it possesses virtually no indigenous energy resources. Japan is thus dependent on the supply of more than 80% of its energy and virtually all of its oil supply from overseas, especially from the Middle East. This makes Japan the most energy insecure state on earth, and, given its importance in the global economy, maximising Japanese energy security remains a significant global issue.

Oil Dependency and the Changing Energy Mix

At present, Japan is the world's second largest crude oil importer after the United States, although China, which became a net oil importer in 1993, is expected to surpass this in the very near future. From the 1960s, the importance of oil as a source of Japanese energy grew rapidly, and from 1960-1970, Japan's oil consumption more than trebled, associated with rapid economic development. However, due to the oil shocks of 1979 and 1985, consumption leveled off somewhat, and, over the past decade, consumption has grown more slowly and is projected to grow relatively slowly over the next decade. However, although Japanese energy policy is aimed at reducing oil dependence from its peak of 77.4% of primary energy supply in 1973 down to 52.9% in 2000, on any scenario, reliance on oil as a source of energy will continue to be in excess of 40% over the next three decades (METI, 2003).

The conscious attempt by the Japanese government to aim to reduce the relative importance of oil as a source of primary energy occurred especially after the second oil shock. Furthermore, environmental demands to reduce greenhouse gases have, in part, meant that the relative importance of coal has remained fairly constant. The main 'beneficiaries' of the relative decline in the importance of oil have been nuclear energy, which is projected to become more important than coal by 2010, and natural gas, the importance of which grew rapidly, especially after the first oil shock.

Maximising Japanese energy security has involved a consideration of policies designed to implement a shift to energy sources which are considered by the government to be of "lower supply risk", both in terms of their susceptibility to sudden price increases but also in terms of the maintenance of large, stable long-term supplies. Coal is seen to be a good energy source, especially in terms of price considerations, but the government feels that, especially from the viewpoint of CO_2 emissions, that the further development and use of natural gas and especially nuclear power is preferred.

For Japanese energy policy-makers, while natural gas is seen to pose lower supply risks than oil, dependence on gas imports, even though that dependence is regional, makes it less attractive than nuclear power in the long-term.

The Nuclear Energy Question

It is a terrible historical irony that nuclear power is likely to assume such an important role in the Japanese energy security profile of the future, particularly in view of its Peace Constitution, the propensity for serious earthquake damage and public concerns over the safety record of some nuclear installations. On the other hand, Japanese government commentators would argue that nuclear power is an excellent source of power from an energy security perspective since uranium offers relative price and supply stability and can almost be seen as a "semi-domestic energy source" due to the prospects for fuel storage and recycling, referred to as "medium-to-long-term energy security". In addition, no CO_2 is emitted in the process of electric power generation in nuclear plants, making it attractive from that particular environmental security standpoint (Irie and Kanda, 2002).

While some countries are planning either to scale down their use of nuclear energy, such as Germany, or not to build any additional nuclear reactors, such as the United States up until recently, at September 2003, Japan had a further three under construction and a further 12 planned. While the USA with 104 of the world's 440 reactors and France with 59 are the world's largest consumers of uranium, Japan presently is the third largest and clearly plans to consume a great deal more over the next decade.

It seems, however, that nuclear energy in Japan presently has something of an image problem. The 2003 Japanese government White Paper on nuclear energy was the first to be published for five years due mainly to serious incidents during that period which fuelled safety concerns. In September 1999, Japan's worst-ever nuclear accident occurred at a uranium processing plant in Ibaraki Prefecture which resulted in more than 650 people being affected by radiation leakages. This accident was followed by a revelation in 2002 that Tokyo Electric Power Company (TEPCO) falsified safety reports on reactor faults, thereby damaging public trust in the nuclear power industry and forcing TEPCO to shut down all 17 of its nuclear reactors for safety checks and repairs (*Japan Times*, 20/12/03). This, in turn, led to a major public campaign to increase energy efficiency and the combination of this, together with high alternative energy import volumes and the relatively cool Japanese summer of 2003 helped avert major power shortfalls (*Japan Times*, 23/11/03). As a result of these issues, in November 2003, the International Energy Agency (IEA) urged Japan to try and restore public confidence in nuclear energy. Its Executive Director was quoted as saying: "It is important for Japan to improve public acceptance of nuclear energy as a way to ensure stable energy supply and mitigate climate change" (*Japan Times*, 19/11/03)

In December 2003, public concern over nuclear safety led to the Tohoku Electric Power Company to abandon plans to build a nuclear plant in Niigata Prefecture, after 30 years of planning, as a result of local opposition to land acquisition (*Japan Times*, 25/12/03).

Japan's Indian Ocean Energy Dependence

Maximising Japan's energy security has inevitably involved a close economic and political relationship with stable long-term energy suppliers, especially from the 1980s. Regional energy dependence upon states which are perceived to be both politically unstable and unwilling to commit to a strong long-term economic and political relationship is important Japanese policy considerations.

In the case of oil, the degree of Japan's dependence upon Middle East suppliers has varied somewhat over the past 40 years, primarily as a result of a combination of oil shocks, variations in domestic economic performance and perceptions of supply risk. Peak dependence was in the mid-to-late 1960s, but both oil shocks brought dependence down below 75%, although, in the last decade, this has been increasing and seems to be moving towards 1960s highs (EDMC, 2003).

From an Indian Ocean perspective, three IOR-ARC member states – United Arab Emirates, Iran and Indonesia – supply Japan with approximately 40% of its oil imports. This fact, together with the overwhelming importance of the oil routes to Japan through the Indian Ocean are two of the main reasons why Japan wished to become an IOR-ARC dialogue partner. Another is linked to the Indian Ocean as being a region of underdevelopment and the global importance of the Japanese ODA programme. From a regional perspective, in 2002, Japan was the largest aid donor to every state which borders the Bay of Bengal (DAC). Both actual and potential Indian Ocean energy suppliers are likely to remain a key component of Japan's energy security strategy for the foreseeable future in order to ensure long-term stable supplies from stable states.

This is one of the principal reasons why the proposal for Japan to develop one of Iran's largest oil fields at Azadegan (with an estimated reserves of 26 billion barrels makes it about the size of the reserves of Mexico) is potentially so significant. The Japanese trade minister had secured preferential negotiating rights on the Azadegan field in November 2000, and, in July 2001, agreement was reached to work cooperatively towards allowing a consortium of Japanese firms to develop Azadegan. Japan had hoped that this oil project could secure a new source of energy to help make up for the loss of rights to the Khafji oil field in Saudi Arabia in 2000 (*Japan Times*, 2/07/03).

However, pressure on Japan from the United States not to go ahead with this development due to the US position over the possibility of Iran's development or possession of nuclear weapons, together with the prospect of Chinese competition for overseas oil resources, created a complex security conundrum for Japanese policymakers. However, the role and importance of the United States in these discussions has certainly been contested within Japan. For example, the Defence Agency Director-General, Shigeru Ishiba, was quoted on Japanese television in August 2003 as saying that: "I do not think that we should give it up because America says so" (*Japan Times*, 25/08/03). On the other hand, there are those who are equally concerned about the resilience of the US-Japan security alliance and thus might be persuaded to side with America. Others are quite cynical about long-term US oil intentions and US relations with Iran. Yet others argue that continued economic interaction and peaceful cooperation and dialogue is the most effective approach in order to "secure oil while keeping the alliance" (*Japan Times*, 11/07/03).

In the event, a contract for the exploitation of Azadegan was signed on 19 February 2004 on behalf of the Japanese consortium and Iran's Deputy Oil Minister. While Japan hoped that the signing of the contract would help further develop Japan-Iran relations, especially if the latter complies fully with IAEA resolutions and NPT obligations, the United States was officially "disappointed" with the outcome.

As the largest net exporter of energy in IOR-ARC and the largest coal exporter in the world, and given Australia's considerable energy reserves, it comes as no surprise that Australia and Japan possess a "long-term energy relationship" (Panda, 1982). The post second world war relationship goes back to the 1957 Agreement on Commerce which coincided with significant coal exports from Australia to Japan. Today, Japan receives more than 45% of all Australian coal, virtually all liquefied natural gas and more than 22% of uranium (Commonwealth of Australia, 1999, 11). South Africa is also another important source of coal for Japan. Indonesia and Malaysia are also important sources of natural gas.

The 1982 Agreement for cooperation in the peaceful uses of nuclear energy was an additional important step for the two 'non-nuclear' states of Australia and Japan. For Japan, Australia supplies about 50% of all of its coal imports and about 20% of all LNG imports. Since Japan's consumption of nuclear energy is planned to increase, and given that Australia possesses the largest known recoverable reserves of uranium in the low cost category (27% compared with Kazakhstan's 17% and Canada's 15%) and given that Australia is presently only the second largest uranium producer, there are very strong prospects for a much closer uranium trade between the two states. Apart from reactor safety issues, the safety of nuclear waste disposal is an important common energy security issue. In the case of Japan, more than 160 shipments of spent nuclear fuel have been sent to the UK and France and recovered fissile materials are returned to Japan as nuclear fuel (World Nuclear Association, 2003a). Clearly, many Indian Ocean states, along with Japan, have an important joint concern over many aspects of energy security, broadly-defined.

Towards an East Asian Energy Strategy

Based principally on the energy security goal of the diversification of energy suppliers, and in part, on notions of proximity associated with the US hemispheric energy supply strategy, Japan is also pursuing an East Asian energy strategy. Three especially important aspects of this strategy – Japan-Russia energy linkages, energy security and energy competition, and a concept for an East Asian energy community – will be considered briefly here.

Japan-Russia Energy Linkages

Since Russia is currently the world's most important energy niche economy as measured by net energy exports, and, given Japan's relative proximity, there has thus been an ongoing dialogue over the complementarity between the resource holder and the capital and technology holder, despite the fact that both Japan and Russia are still technically in a state of war and are in dispute over the Kurile islands.

One potential long-term supply of natural gas is via the proposed gas pipeline linking Tokyo with Russia's Sakhalin

Island, which is part of the so-called Sakhalin I oil and gas development project undertaken by ExxonMobil, Sakhalin Oil and Gas Development Company and a number of Russian companies. Under this major project, natural gas production was expected to begin in 2008 with the 2,400 kilometre pipeline planned to deliver gas to the Tokyo metropolitan area. However, TEPCO, which would be the biggest user of the gas, is concerned about possible oversupply, while others believe that, if Japan decides to reduce its dependence on nuclear power as a result of concerns over the safety of nuclear plants, then demand for gas could substantially increase (*Japan Times*, 1/10/03).

Another project, Sakhalin II, which is run by a consortium comprising Royal Dutch-Shell Group and Japan's Mitsui and Mitsubishi, has already signed contracts with Tokyo Gas and Tokyo Electric Power Company to supply liquefied natural gas from Sakhalin Island to Tokyo from 2007. However, a coalition of environmental NGOs opposes the Sakhalin II project because of potential dangers to the island's ecosystem and to the Western Pacific gray whale and that the consortium has neglected anti-seismic norms in the design of the project in order to cut costs (*Japan Times*, 3/11/03).

Energy Security and Energy Competition

For at least a decade, Eastern Siberia has been a potential target for Japanese energy investment, but the prospect of a 4,000 kilometre Japan-financed oil pipeline route linking Angarsk with Nakhodka has been complicated somewhat by Chinese competition and by a Chinese proposal for a shorter 2,400 kilometre route from Angarsk to Daqing in China's Heilongjiang Province (*Japan Times*, 13/July/03). The Nakhodka route would enable Russian oil exports not only to go to Japan, but also to South Korea, Taiwan and even the United States (*Japan Times*, 5/08/03).

The competition for Siberian oil surfaced in May 2003 when one of Russia's large oil companies, Yukos, signed a 25-year agreement with China National Petroleum Corporation to supply oil through the shorter route to Daqing beginning in 2005. In June, however, a Japanese delegation led by the Japanese Foreign Minister Kawaguchi and former Prime Minister Mori visited Moscow to urge the alternative longer route. It seemed that high-ranking Russian officials, including some close to the Russian President, preferred the Japanese position on this issue. A month later, Yukos came under Russian legal scrutiny and charges of tax evasion and embezzlement were laid, its major shareholders were arrested, and charges were also brought against Mikhail Khodorkovsky who had been providing donations to political groups opposed to the Russian President. The United States position on this issue is that they would prefer Japan to secure oil from Siberia than from Iran (*Japan Times*, 28/08/03). Although a final decision on the oil pipeline route was expected by the end of 2003, it has been apparently delayed due to the considerable environmental assessment work required for the project (*Japan Times*, 5/09/03). Apart from environmental concerns, in order to be economically viable it seems that the longer pipeline would need to carry at least 50 million tons of crude oil a year compared with the shorter route volume of 20 million. It has been suggested by one commentator that there is insufficient oil to justify the longer route (*Japan Times*, 17/09/03).

Meanwhile, although China appears to remain confident that its route will prevail (*Japan Times*, 22/09/03), the Russian Prime Minister has stated that the pipeline deal with China has been "postponed" while at the same time stressing that such decisions are in the hands of private companies beyond the direct influence of the Russian government (*Japan Times*, 26/09/03). Japan, on the other hand, had hoped that a decision in its favour would result from the Russian Prime Minister's visit to Tokyo in December 2003, although the decision will likely be delayed until President Putin's visit to Japan some time in 2005.

Towards an East Asian Energy Community

While energy competition among the world's largest energy users is unsurprising, a much more attractive option aimed at maximising the energy security of all is to create an energy community. To some degree, the concept of an Asia-Pacific energy community has already gained some currency (Miyakawa, 1996, 73).

Concerns over energy led to a call by the 2001 Shanghai APEC summit meeting for member states to more effectively cooperate to ensure the maximisation of regional energy security.

In short, a need was recognised for the introduction of new energy security measures that would involve multilateral rather than bilateral thinking. Such a new approach in East Asia, for example, might involve better cooperation and coordination of energy policy among regional states and even the creation of an East Asian Energy Agency (EAEA). It could lead to the construction of a Northeast Asian gas pipeline which would also help maximise environmental security. Cooperation in the development of nuclear power could enhance regional stability and environmental security since the United States bans the export of nuclear power equipment to China. It might also lead to the creation of a common Japan-Korea energy market as part of a free trade agreement (Toichi, 2003).

Energy Security and Environmental Security

The relationship between energy use and CO_2 emissions is well known and well-documented. What this means, of course, is that the heaviest energy consumers per head of population are likely to be the heaviest CO_2 emitters per head of population. Interestingly, when the per capita energy use is greater than 5mtoe there is a disproportionate increase in emissions per head. The North American states possess the largest CO_2 emissions per head in the world (Table 8.1).

In an absolute sense, Japan has the world's fourth largest CO_2 emissions, although its per capita emissions indicate a degree of environmental efficiency based both on environmental technology and a larger percentage of nuclear power in the energy mix. States such as France with a relatively high nuclear energy dependence have much lower absolute and per capita CO_2 emissions (Table 8.1)

In Kyoto Protocol discussions, Japan agreed to a target of 6% reduction in its CO_2 emissions below 1990 levels. It has been estimated that for Japan to reach this target by 2010 would result in increased costs of production in its carbon intensive sector (especially iron and steel and electricity) which would result in lost competitiveness and reduced outputs which in turn would reduce demand for fossil fuels, especially coal. The outcome of this could be that by 2010 Japan could reduce its coal imports by 52%, its oil imports by 12% and its LNG by 1.5%. For Australia, this could mean a reduction in the order of 55% of coal imports (Commonwealth of Australia, 1999, 6).

The introduction of a carbon tax might be one way of allowing the market to dictate the reduction in CO_2 emissions. However, the Japan Business Federation (Nippon Keidanren) is seemingly opposed to such a policy objective (*Japan Times*, 19/11/03).

Clearly, although a switch away from coal towards natural gas would lead to reduced CO_2 emissions, the nuclear industry tries hard to market *its* product as being ecologically sustainable (World Nuclear Association, 2003b) even though it is potentially a considerable threat to environmental security. Apart from the problem of nuclear waste, it has been estimated that a large-scale radiation leak in Japan could kill in excess of 400,000 people and cost up to ¥460 trillion over 50 years (*Japan Times*, 28/October/03).

Table 8.1. Carbon dioxide emissions 2000 (mt CO_2).

Country	Absolute	Per Capita
USA	5665.44	20.57
China	2996.77	2.37
Russia	1505.74	10.34
Japan	1154.84	9.10
India	937.28	0.92
Germany	832.95	10.14
Canada	526.77	17.13
Korea	433.57	9.17
Italy	425.73	7.37
France	373.26	6.18

Source: IEA, 2002, pp. 48-57

Increased energy efficiency, associated with environmental technology and the increased use of alternative renewable sources of energy. would be more appropriate methods of enhancing environmental security. However, it has been calculated that the cost of reducing CO_2 emissions in Japan by using bioethanol fuel would be more than 50 times greater compared with purchasing CO_2 emission credits, as allowed under the Kyoto Protocol (*Japan Times*, 18/11/03).

From a broader East Asian perspective, energy conservation is essential for the enhancement of energy security. On this basis, the Japanese Ministry of Economy, Trade and Industry (METI) is undertaking a Green Aid Plan (GAP) designed to improve the level of bilateral cooperation towards solving energy-related environmental issues with seven Asian states – China, India, Indonesia, Malaysia, Philippines, Thailand and Vietnam. In particular, GAP aims to provide assistance for the creation of a system for energy conservation primarily through education and legislative advice.

Japan's Energy Security Future

To maximise its energy security Japan needs both to reduce its energy dependence on fossil fuels and to increase its reliance on states which are able to provide long-term supplies and are stable and willing to enter into long-term economic and political relationships. Developing strong relations with new energy regions such as the Caspian Basin is also an important consideration. Furthermore, accessing new energy supplies hemispherically is an essential energy policy goal along with the possible development of a regional energy community.

Even though Japan itself has few energy resources of its own, this may not be a permanent situation. The Japanese government had already confirmed the existence of a natural gas field off the Aomori coast in 1999, and, in late 2003, the Japan Energy Corporation began exploring for natural gas in that region as part of a consortium (*Japan Times*, 23/10/03). Furthermore, the government believes that it can claim up to 650,000 sq km of continental shelf as part of its EEZ, which would be 70% larger than the country's land area. Large-scale exploration planned for this EEZ, including a search for non-conventional energy sources, is a high priority over the next decade.

In addition, increased energy conservation and the move towards alternative renewable energy sources will enhance environmental security. In 2003, the USA, Japan and 13 other states agreed to "The International Partnership for the Hydrogen Economy" to collectively develop hydrogen and fuel-cell technologies. Toyota Motor Corporation, as well as others in Japan and elsewhere, is keenly interested in the introduction of hydrogen fuel cells (*Japan Times*, 22/11/03). Wind power has also seen a rapid growth in Japan over the past two years.

Perhaps the most geopolitically sensitive alternative energy security proposal is Japan's involvement in the International Thermonuclear Experimental Reactor (ITER) proposal for nuclear fusion along with China, EU, Russia, South Korea and the United States. The outcome would be the generation of large quantities of cheap power with water as its main by-product. The consortium of 6 states has been discussing the prospect of siting a US$38 billion nuclear fusion reactor somewhere in the world, although there is disagreement as to where this might be placed. China and Russia, for example, favour locating the reactor in Cadarache near Marseille in France. However, this is opposed by the United States and South Korea who favour a location at Rokkasho, a small fishing port close to the US military base in northern Japan. On the one hand, France sees the opposition to Cadarache as punishment for not supporting the war on Iraq. On the other hand, China is concerned about the location of the reactor at Rokkasho since it would turn Japan into a virtual thermonuclear superpower and provide an ongoing supply of materials necessary to build advanced nuclear weapons. However, it is possible that, if no agreement can be reached, the EU would press ahead and build the reactor in Cadarache.

Conclusion

While Japan remains dependent on oil as its major source of energy, maintaining a national oil reserve and the stockpiling of oil at strategic locations around Japan becomes an essential element of Japanese energy security. However, for the future, it has been argued that, in the 21st Century, both the end of the Cold War paradigm and increasing globalisation demand a reevaluation of Japan's energy diplomacy. It has been suggested that three considerations should help to guide such a process. First, it has become even more critical to better coordinate security policy and energy policy. Second, it is important for Japan to become more influential among oil-producing states in the broader Indian Ocean region, especially in the Middle East, and to develop better cooperative relations with Northeast Asian states.

Third, much closer coordination is necessary between energy diplomacy and environmental diplomacy (Toichi, 2003, 50).

Acknowledgement

Much of the background research for this Chapter was undertaken during a Visiting Research Professorship at Kyoto University in the latter part of 2003. I am extremely grateful to the University and especially to Professor Akihiro Kinda and to all of his colleagues in the Department of Geography for making this visit so stimulating, enjoyable and productive.

References

Awad, S. (1999), 'The political economy of oil', *Third World Quarterly*, Vol. 20 (6), pp. 1221-1225.

Bahgat, G. (2001), 'United States energy security', *The Journal of Social, Political and Economic Studies*, Vol. 26 (3), pp. 515-542.

Balmaceda, M. M. (1998), 'Gas, oil and the linkages between domestic and foreign policies: the case of Ukraine', *Europe-Asia Studies*, Vol. 50 (2), pp. 257-286.

Belgrave, R., Ebinger, C. K. and Okino, H. (1987), *Energy Security to 2000* (Aldershot: Gower).

Bohi, D. R. and Montgomery, W. D. (1982), *Oil Prices, Energy Security and Import Policy* (Baltimore: Resources for the Future).

BP (2003), *Statistical Review of World Energy*, 52nd edition.

Calder, K. E. (1996), 'Asia's empty tank', *Foreign Affairs*, Vol. 75 (2), pp. 55-69.

Clawson, P. L. (1995), *Energy and National Security in the 21st Century* (National Defense University Press).

Commonwealth of Australia (1999), *Japan's Energy Future: Dilemmas, Policy Targets and Practicalities*, a report by an industry-government energy delegation to Japan, Canberra.

Cutler, R. M. (1999), 'Cooperative energy security in the Caspian region: a new paradigm for sustainable development?', *Global Governance*, Vol. 5 (2), pp. 251-271.

DAC website: http://www.oecd.org.countrylist - Development Assistance Committee, OECD.

Drennen, T. E. and Erickson, J. D. (1998), 'Who will fuel China?', *Science*, Vol. 279, page 1483.

Energy Data and Modeling Center (EDMC) (2003), *2003 EDMC Handbook of Energy and Economic Statistics in Japan*.

Fesharaki, F. (1999), 'Energy and the Asian security nexus', *Journal of International Affairs*, Vol. 53 (1), pp. 85-99.

Government of Japan (2001), *Comprehensive Review of Japanese Energy Policy*, (Tokyo: ANRE, METI).

International Energy Agency (IEA) (2000), *China's Worldwide Quest for Energy Security* (Paris: OECD).

International Energy Agency (2001), *Oil Supply Security: The Emergency Response Potential of IEA Countries in 2000* (Paris: OECD).

International Energy Agency (2002), *Energy Efficiency Update: Japan* (Paris: OECD).

International Energy Agency (2003), *Key World Energy Statistics* (Paris: OECD).

Irie, K. and Kanda, K. (2002), 'Evaluation of nuclear energy in the context of energy security: analysis based upon Japan's policy documents', METI, Japan Energy Policy Institute.

Japan Times, various issues.

Keith, R. C. (1986), *Energy, Security and Economic Development in East Asia* (London: Croom Helm).

Kelly, D. (2002), 'Rice, oil and the atom: a study of the role of key natural resources in the security and development of Japan', unpublished paper.

Manning, R. A. (2000), *The Asian Energy Factor: Myths and Dilemmas of Energy, Security and the Pacific Future* (London: Palgrave).

Marples, D. R. and Young, M. J., eds. (1997), *Nuclear Energy and Security in the Former Soviet Union* (Boulder: Westview Press).

Martin, W. F., Imai, R. and Steeg, H. (1996), *Maintaining Energy Security in a Global Context* (Brookings, The Trilateral Commission).

METI (2003), 'Energy in Japan', http://www.meti.go.jp

Miyakawa, Y. (1996), 'Mutation of international politico-economic structure and the development of the Pacific maritime corridor in the East Asia orbit', in Rumley, D., Chiba, T., Takagi, A. and Fukushima, Y., eds., *Global Geopolitical Change and the Asia-Pacific: A Regional Perspective* (Aldershot: Ashgate), pp. 49-75.

Morse, E. L. and James, R. (2002), 'The battle for energy dominance', *Foreign Affairs*, Vol. 81 (2), pp. 16-31.

Morse, R. A., ed. (1981), *The Politics of Japan's Energy Strategy: Resources-Diplomacy-Security* (University of California: Institute of East Asian Studies).

Panda, R. (1982), *Pacific Partnership: Japan-Australia Resource Diplomacy* (Rohtak: Manthan).

Rowley, A. (1991), 'Naked power: Japan looks to reduce its dependency on oil imports', *Far Eastern Economic Review*, 12 December, pp. 52-3.

Saito, M. (2000), *The Japanese Economy* (Singapore: World Scientific Publishing).

Soeya, Y. (1998), 'Japan: normative constraints versus structural imperatives', in Alagappa, M., ed., *Asian Security Pratice: Material and Ideational Influences* (Stanford University Press), pp. 198-233.

Sugihara, K. and Allen, J. A., eds. (1993), *Japan in the Contemporary Middle East* (London: Routledge).

Toichi, T. (2002), 'Japan's energy policy and its implications for the economy', paper delivered to the 3rd Japan-Saudi Business Council Joint Meeting, Riyadh, March.

Toichi, T. (2003), 'Energy security in Asia and Japanese policy', *Asia-Pacific Review*, Vol. 10 (1), pp. 44-51.

Toman, M. A. (2002), 'International oil security problems and policies', *The Brookings Review*, Vol. 20 (2), pp. 20-23.

Tsuru, S. (1993), *Japan's Capitalism: Creative Defeat and Beyond* (Cambridge University Press).

World Nuclear Association (2003a), 'Japanese waste and MOX shipments from Europe', web site: http://www.world-nuclear.org/info

World Nuclear Association (2003b), 'Sustainable energy', web site: http://www.world-nuclear.org/info

Yergin, D. (1988), 'Energy security in the 1990s', *Foreign Affairs*, Vol. 67 (1), pp. 110-132.

CHAPTER 9

China's Expanding Energy Deficit: Security Implications for the Indian Ocean Region

Swaran Singh

According to China's Tenth Five Year Plan 2001-2005, by the year 2020, China is estimated to import about 500 million tonnes of oil and 100 billion cubic meters of gas to cater for its ever-growing domestic demand. This would mean that the share of China's imports in its domestic consumption of oil and gas will rise to become 72 per cent and 50 per cent respectively of China's total consumption at that time.[1] More conservative western estimates project China's energy imports for 2020 to be 30 per cent for gas and 60 per cent for oil.[2] Either way, given China's recent turbulent history of the last 55 years, such projections remain highly discomforting for China's new leadership. This is especially the case in view of China's continued concerns about Western powers' policies of containment in the past and their continued skepticism towards China's rise. The fact is that China has since moved from once being East Asia's top oil exporter during the mid-1980s to East Asia's second largest importer of oil (after Japan) since 1993. This ever-expanding energy deficit has come to be seen as the most serious threat to China's continued economic development, on which hinges its political stability and social cohesion.

Then there are general concerns about the forces of globalization and China's recent entry into the World Trade

Organisation (WTO) that makes China far more vulnerable and exposed to global energy trends. With global energy consumption expected to rise by about 60 per cent by 2020, China, with its expanding energy deficit, will be increasingly drawn into an intense global competition for energy resources. China's attempts to import energy will face severe competition in its immediate region dotted by other high-energy consumption countries. Especially in the case of oil – where the total world production by 2020 is likely to reach 4 billion tonnes per day and the total share available oil for exports will be 1.5 billion tonnes – China would have to procure the bulk of international oil exports to provide for its domestic needs. All of this has obvious security implications for China's periphery and especially for countries that have had difficult relations with Beijing.

It is in this evolving context of China's growing energy demand and its expanding stake in global energy markets that this chapter tries to examine the nature and magnitude of China's expanding energy deficit and to highlight its security implications for the Indian Ocean Region (IOR). The IOR has lately emerged as the new hub of energy reserves and supply lines and has lately witnessed foreign interventions of all kinds. Indeed, given its physical proximity, recent years have already witnessed a heightened Chinese activism in the Indian Ocean rim, and, with India's energy needs expanding on similar lines, New Delhi has already begun to be concerned over the security implications of China's ever-expanding energy deficit.

Debates about China's Energy Deficit

There have been varying explanations of the nature of China's energy deficit. Indeed, Chinese experts themselves remain divided about several fundamental questions. Chinese scholars, in general, like to emphasise that what makes China's energy deficit a critical problem is not a simple situation of a "shortage" of energy. Indeed, even today China continues to export energy including certain kinds of crude and refined oil. Instead, it is China's ever-growing demand for "clean" and "efficient" energy – especially oil – that lies at the core of this crisis in-the-making.[3] For the Chinese, this remains primarily a structural adjustment problem where coordination and vision have been found wanting. China remains, therefore, 'a typical example of a resource-rich yet import-dependent country because its energy policy has not kept pace with developments in the changing nature of China's energy needs. Otherwise, China is not short of energy, even in those "clean" and "efficient" energy sectors. Even in a clean and efficient energy source like gas, for example, China possesses abundant reserves of 53.3 trillion cubic feet (as of March 2003).

There also remain inconclusive debates on why and how the energy deficit might become a security threat in international relations. Once again, Chinese experts believe that it is the nature of international politics that makes energy securitized. One author describes oil as "an important weapon of diplomacy, the blood of industry, life of economy, a guard against aggression, and, therefore, target of big powers."[4] As a result, the security implications for the Indian Ocean region are not seen as flowing exclusively from China's energy deficit but from the fact that much of China's oil imports flow through Indian Ocean sea lanes with much passing regions ridden with turbulent politics. For example, the nature of threat projections around the scramble for oil has more to do with events like the two Gulf Wars that clearly overplay energy becoming the strongest flashpoint in the Asia-Pacific region. Of course, the rising demand for oil imports will only exacerbate these political equations.

Besides, all of these energy deficit projections completely neglect the other parallel trends that also impact on developments in the energy sector and might complicate China's attempts to deal with its energy deficit. Any discussion on the implications for the Indian Ocean region, therefore – that is expected to become the focus of energy deficit-related conflicts – must emphasize the fact that, in addition to being impacted by China's growing energy deficit, the nature of its security implications depends on the nature of inter-state ties. This region, that has world's busiest sea lanes of communication (and potentially some pipelines) for transferring energy and other materials and services, may have other triggers to an energy crisis as well, and these will continue to guide China's genuine interest in the Indian Ocean. Accordingly, China's growing stakes and participation in the Indian Ocean region will have to be viewed in this larger perspective of two-way traffic of multifaceted perceptions and interactions in several sectors in which the energy deficit only forms a part. The fact is that both foreign trade and energy have been the main locomotives of China's successful modernization of the last two decades and China has lately focused on expanding its engagement with Indian Ocean countries.[5]

China's Opening Up and Engagement

Going by China's recent history since Deng Xiaoping launched economic reforms, China's energy sector has both *necessitated* and has also *facilitated* China's engagement with the Indian Ocean rim. The picture was exactly the opposite when Mao ruled China during its first 27 years. Throughout this period, China was generally aloof and neutral to most outside developments in the energy sector. This was possible because of China's policy of seeking "self-sufficiency" (or *zili gengsheng*). This was viewed in terms of exploring and using China's domestic resources and capabilities to "keep the initiatives in one's own hands".[6] Although China had begun to export energy from the mid-1970s, yet this was too modest to have either a positive or negative impact on China's then ideologically-driven policy responses to regional and global developments. This was so even with regard to developments in the global energy sector, and events like price fluctuations or supply disruptions made no difference to Beijing. China had little response even to historic events like the cartelization of oil during early 1970 which resulted in oil prices going up four times, resulting in major downsizing of development projects in the oil-importing world and to major diplomatic footwork and rhetoric by the major powers.

It was only following China's rapid economic development during the 1980s and 1990s that China developed a stake in the global energy market. To begin with, China's opening up and reforms from the early 1980s were to witness a major reorganisation of its energy sector during 1988.[7] This re-orientation of China's energy sector was to be further hastened following China becoming a net importer of oil from 1993 (Figure 9.1). All of this has also been further exacerbated by China's domestic production failing to keep pace with the qualitative change in its rising domestic consumption. Between 1990 and 2000, for example, China's oil consumption doubled from 2.1 million barrel per day to 4.6 million barrels per day. The future projections of China's oil demands for 2020 now range from 6.4 million barrels per day to 10.1 million barrels per day.[8]

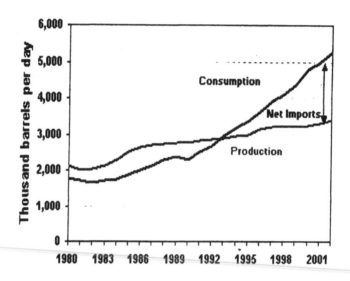

Figure 9.1. China's oil production and consumption, 1980-2002
(Source: EIA).

This has since forced China to devise newer strategies to open up its energy sector and to focus on engagement with global players in different sectors of energy. As a result, China's energy deficit has since evolved newer connotations and its oil deficit remains pregnant with possibilities of becoming another ping pong ball in China's international relations. Besides, the debate on widening security has also since evolved a strong new thesis where energy is to become increasingly securitised and energy flashpoints are to become the focus of international conflicts in the coming years.[9]

As a result, in addition to its domestic sector, China's energy policy has since begun to focus on its offshore energy assets, leasing foreign oilfields and acquiring stakes in foreign energy establishments and joint ventures as new instruments for ensuring energy supplies. In addition, energy sector has also since emerged as an integral part of China's overall maritime thinking and its naval assets have become integral to China's holistic maritime vision for the 21st century.[10] Whereas China's conventional energy strategies had focussed on (a) self-sufficiency in terms of exploiting only domestic resources, (b) emphasising conservation and the efficient use of energy sources, and (c) diversifying both the sources of energy and also the sectors of energy to maintain a reliable balance, China's newer energy strategy focus from the early 1990s has shifted towards "cooperation" and "engagement" in terms of joint development, foreign direct investment and technology transfers as well as an innovative approach towards alternative and renewable energy sources. For example, China's natural gas consumption that was 0.5 Trillion Cubic Feet (Tfc) for 1990 is projected to reach to 6.1 Tfc by 2025. This involves an increase at the rate of 1120 per cent, whereas global consumption is projected to rise only by 140 per cent for the same period.

China has also experimented with the privatisation of its best-performing firms in the energy sector. These were opened up for collaboration with major players in the field as well as for stakeholding with the initial public offerings (IPOs) respectively during 2000 and 2002.[11] This has since become a routine in China's energy sector. Indeed, signed after 12 years of negotiations in 2000, a $4.35 billion petrochemical deal between China National Offshore Oil Corporation (CNOOC) and the Royal Dutch/Shell for Huizhou (in Guangdong) remains China's single largest joint-venture and the world's leading petrochemical project so far.[12] This clearly shows the position of energy in China's list of priorities. Indeed, such collaborations have lately come to be criticized for having made China, (a) increasingly dependent on foreign suppliers and on resources which are exhaustible, (b) getting into serious competition with other major oil and gas importing powers, and, (c) becoming more concerned with ensuring the safety of supply lines, thereby multiplying its stakes in global stability, security and order.[13]

Once again, a large number of Chinese experts continue to caution that the security implications of these new trends remain far too exaggerated. Most Chinese studies try to explain how the problems of China's energy deficit are all very much manageable and that these are not necessarily expected to lead to an international crisis. Nevertheless, studies focusing on the short and medium-term seem inclined to accept that China has failed to keep pace with increasing demand for providing "clean" and "efficient" energy; that this, if continued, the trend is very likely to have serious security implications for both China's domestic politics and for its foreign relations. China's expanding deficit in oil remains the one that seems the least manageable of all. Though there may not be a visible "energy-security-crisis" in sight at present, yet the growing imbalance in China's domestic demand for oil seems to have risen sharply (Figure 9.1). Apart from oil reserves, the problem also lies with the quality of China's outdated oil refining technologies that lie at the root of its expanding oil deficit. Further, China seems to have done precious little to either expand its gas production or develop its alternative and renewable energy sources to mitigate security implications from its growing and threatening oil deficit. All of this makes China's energy deficit a much complex problem than what can be appreciated from a simpler supply-demand analysis.

Deficit Defined I: Structural Contradictions

It needs to be underlined that China's energy deficit cannot be understood in stark and simple terms, as may be the case with Japan or South Korea, which are other two major importers of energy through the Indian Ocean rim. China is not really an import-dependent state in the same manner. China's energy deficit is not one of falling short on reserves or even supply of energy compared to its total national demand. China indeed prides itself of having vast coal and gas reserves as well as a great nuclear and hydropower potential, and the later two have since become China's policy priorities. In addition, there always remain renewable sources of energy for which again China does not feel to be placed in any disadvantageous position. Given its vast territory, China seems particularly suited to the exploration and use of these renewable and alternate energy resources. The question of China's energy deficit, therefore, remains far more subtle and nuanced in terms of the structural contradictions and a flawed energy policy that remains too complicated to be fixed by any easy *ad hoc* policies.

In brief, China's energy deficit results form the rigidity and inflexibility of China's energy strategies where the bulk of production and distribution channels as well as its processes and techniques have not been able to transform themselves to keep pace with the increasing demands for new kinds of cleaner and efficient energy which has been the most visible trend in China since early 1990s. There is, of course, a physical location problem as well. The bulk of China's energy resources remain located in its western and north-central regions, while the bulk of its consumption remains concentrated in its developed eastern coastal regions. The rapid development in China's east has witnessed not only an equally rapid rise in purchasing power and demand for energy but this has been accompanied by an increasing awareness about global trends in energy, resulting in demand for obtaining what is not produced domestically, from outside. This compels China to engage and compete with outside market forces and makes it vulnerable to global price fluctuations and supply disruptions which often become both cause as well as consequence of China's political and security equations within and with other countries.

To illustrate these structural contradictions, though the share of demand for pollution-causing coal continues to fall gradually, yet coal still accounts for a significant 67 per cent of China's total energy consumption, whereas the international average share of coal in global energy consumption remains as low as 24.37 per cent.[14] By contrast, in terms of the share of cleaner energy sources such as natural gas – of which China has abundant reserves – the global share stands at 24 per cent of total global energy consumption (and even for Asia the average share remains above 8 per cent), for China it remains a mere 2.5 per cent.[15] China's electricity production – another most clean energy source – also remains dependent on coal. Electricity production in China comprises 82.3 per cent thermal (Coal) power, 16.4 per cent hydropower and only 1.2 per cent nuclear power.[16] Rising awareness and norms on environmental pollution have since become a major hurdle in sustaining the pace in much of the outmoded parts of China's energy sector. As a result, while a 6.3 per cent growth in China's electricity sector did manage to keep a balance between demand and supply during China's Ninth Five Year Plan (1993-1998), during the Tenth Five Year Plan (2001-2005) the production growth rate is expected to fall to about 5 per cent while the demand for electricity is likely to rise much faster.[17]

China's growing awareness and legislation about environmental protection have also come to gradually circumscribe its energy sector.[18] Despite its largest coal reserves and largest hydropower and nuclear potential, China's antiquated technologies continue to create problems in any desire for rapid change. China today finds it too expensive to transform its energy sector technologies in order to conform to all of the prescriptions of the "sustainable development" thesis.[19] And now, China's accession to the WTO is going to further expose its vintage energy sector to the challenge of the world's best energy exploration, production and distribution technologies and management skills, thereby increasing its dependence on foreign collaboration both for importing clean and efficient energy as well as for advanced technologies for joint production and distribution. Without engaging these external players, China's geological and physical advantages, vintage technologies and limited management skills are going to remain a serious roadblock to ensuring a continued flow of energy for both stable and rapid development in its economy and society. It is these structural contradictions in China's energy sector that determine the security implications both for China and its neighbours, including the Indian Ocean rim.

Deficit Defined II: Surplus vs. Deficit

Indeed, when it comes to critical sources of energy like gas and coal and even oil reserves, China presents a case of energy surplus and energy deficit at the same time. The security implications of such a complex energy profile can be far more lasting, imperceptible, widespread and deep-rooted. Apparently, coal continues to be the backbone of China's energy supplies. However, several smaller and inefficient coal mines had to be closed during the 1990s to deal with problems of surplus of certain kinds of not-so-clean coal, although inefficiency was also a factor in their closure. This clearly highlights how structural problems can lead to a situation of a coexistence of a surplus of one kind and a deficit of another kind. This situation has the potential to result in a security crisis, especially if the structural problems are not rectified in time. Indeed, even when they may be corrected in due course of time, a balanced energy sector is still likely to become the flashpoint given its increasing influence on global trends in both rising demand and depleting supplies in gas and especially oil. Ideally speaking, China (and most other industrialised countries) will have to rearrange their overall energy profiles to ensure their continued economic development over a longer period of time.

To cite some comparisons where China has been trying to exploit its technical potential and natural reserves, natural gas presents one example where the share of natural gas consumption in China stands at mere 2.5 per cent of its total energy consumption against the global average of 24 per cent. Interestingly, however, as much as 75 per cent of China's current natural gas consumption continues to be in its industrial sector and the consumption of natural gas by China's civil society remains yet on the margins. This is one area where China can make rapid progress in correcting the structural imbalance in its energy consumption pattern. In nuclear power as well, China's nuclear-generated electricity amounts to only about 1 per cent of its total energy consumption. Despite recent enthusiasm, China's nuclear power remains extremely low compared to the global average of 17.1 per cent. Particularly compared with other nuclear weapons states – 77 per cent for France, 28 per cent for Britain, and 19 per cent for the USA – China seems yet to have tremendous potential which remains to be exploited towards producing clean and efficient energy sources (assuming, of course, in the case of nuclear power, that safety and waste problems can be overcome).

China's ever-increasing trade surplus is another factor that has critical linkages with its expanding energy deficit. The success of China's four modernisations drive has since led to a growing interest beyond its territorial boundaries. Firstly, China's engagement and interdependence with the outside world has been driven by its booming foreign trade that had grown to $851 billion by 2003, with China's foreign exchange reserves reaching a record $403 billon for that year.[20] This level of foreign trade has generated tremendous sea-faring endeavours towards obtaining access to sea lanes of communication and ocean resources. As a result, with its 26 shipyards, China has since become the fourth largest shipbuilder from once being 16th until the mid-1980s. This has had a direct impact on China's Indian Ocean policy. To begin with, this clearly shows in the change in China's policy towards the

South China Sea where projections about its energy potential – that vary between 1 to 17.7 billion tonnes of oil – have since come to define China's preoccupations with this water body.[21] The increasing Chinese control over the South China Sea is seen to provide it with added advantage in extending its operational reach into the Indian Ocean rim.[22] China's forays into Indian Ocean, therefore, need to be examined from two perspectives: (a) a need-based and (b) an ambition-based perspective. This should help in highlighting both the threats and the limitations that lie behind China's engagement with the Indian Ocean rim and also in gauging the security implications of China's expanding energy deficit for the Indian Ocean region.

China's Stakes in the Indian Ocean

In terms of China's increasing stake in the Indian Ocean, recent years have witnessed evidence that clearly points in that direction. In the past as well, China tried to access the Indian Ocean via the Karakoram Highway from the 1950s and then through Burma from the 1990s. However, thanks to China's growing trade and increasing demand for energy in the recent period, China's new engagement has developed several new partnerships in the Indian Ocean rim – for example, South Africa. The Soviet entry into Afghanistan from the late 1970s was another major factor igniting China's indulgence with this region, witnessing an increase in its port visits to the Indian Ocean rim.[23] As a result of China's growing interest and engagement with the Indian Ocean rim, the Council of Ministers of the Indian Ocean Rim Association for Regional Cooperation (IOR-ARC) granted China the status of a Dialogue Member at its January 2000 special session in Oman. Given the fact that China is not a littoral state of the Indian Ocean, this decision by IOR-ARC clearly reflected the importance that member states attach to China's interest and initiatives, track record as well as its growing stake in sea lanes of communication and resources of the Indian Ocean region.[24]

Very briefly, there exist at least five major factors that largely guide and determine China's increasing engagement with the Indian Ocean rim countries. Firstly, with the Atlantic and the Pacific oceans already under the grip of Western powers, this leaves only the Indian Ocean that happens to be surrounded largely by developing and not-so-powerful states. In addition, the Indian Ocean contains the world's largest energy deposits. The Chinese perceive a kind of a power vacuum in this region and have been trying to step into it. Given their increasing engagement and acceptance in the Indian Ocean rim, the apparent Chinese success in this exercise has only further encouraged Beijing's initiatives.

Secondly, China's booming foreign trade routes and especially its expanding energy supply lines remain dotted through the Indian Ocean. As of 2001, China's imports of oil stood at 56.2 per cent from the Middle East, 22.5 per cent from African countries, 14.4 per cent from the Asia-Pacific and 6.9 per cent from Europe and Central Asian countries.[25] Given the fact that China's continued development remains its essential imperative to ensure its stability and peace, and, since China's economic development remains premised on the success of its foreign trade and energy supplies, China is bound to engage with Indian Ocean rim countries.

Thirdly, unlike the Atlantic and Pacific Oceans, the Indian Ocean happens to be one of the most rapidly growing but also tumultuous regions. Furthermore, the evolving new phase of regionalism and multilateralism in the Indian Ocean rim remains in tune with China's evolving new policy orientation. Lately, international terrorism seems also to be concentrated around this region that makes it potentially very volatile. This creates all of the ingredients where the growing interest of big powers like China (in seeking access and influence) can make it an energy flashpoint. Accordingly, the new leadership in China has already expanded its thesis of seeking peace on China's borders to seeking peace in border regions with its focus on the expanded periphery such as the Indian Ocean.

Fourthly, China, having successfully experimented with leasing oilfields in Latin America, is likely to explore new opportunities in the Indian Ocean rim. Accordingly, China's growing engagement with Indian Ocean littoral countries as well as IOR-ARC vindicates its increasing interest in this Ocean. China currently has energy cooperation that spans from Argentina to Bangladesh, Canada, Colombia, Ecuador, Indonesia, Iran, Iraq, Kazakhstan, Malaysia, Mexico, Mongolia, Myanmar, Nigeria, Pakistan, Papua New Guinea, Peru, Russia, Sudan, Thailand,

Turkmenistan, Venezuela and the United States with its obvious linkage to the Indian Ocean rim.

And, finally, this means that even a benign China will remain in competition for other rising states in its periphery and this competition for limited and depleting energy and other supplies will take place increasingly in the Indian Ocean rim. This means that states of the Indian Ocean littoral (and booming new energy consumers like China and India, in particular) must begin to take into account the implications of their increasing energy deficits. This perception of growing competition has already seen China gearing up to deal with other regional players both through engagement and power projections. All of this is likely to have a major impact on Indian Ocean regional security and peace.

China, however, cannot be labeled as the only initiator of this new energy competition. In some ways, China is also going to be a victim of this ever-expanding energy competition in this larger region where just four economies – China, Japan, South Korea and Taiwan – contribute to the bulk of energy consumption. In particular, oil imports of most of these states have been increasing rapidly and carry sinews of an energy crisis and even conflict. Thanks to increasing oil imports by new importing countries, Japan's oil imports, that accounted for 77 per cent of total imports by Asian states in 1992, are projected to shrink to 37 per cent by year 2010. This is because in addition to rising oil imports by parts of Greater China and India, other oil exporters like Indonesia, Malaysia, Brunei and Vietnam are all expected to become net importers by 2010. All of this clearly portends an impending energy crisis in-the-making. China, indeed, has already taken steps and the focus of China's oil diplomacy has since shifted to ensuring greater foreign involvement (FDI and technology transfers) as well as to evolving joint strategies towards ensuring the safety of SLOCS and against any misuse of Ocean choke points.

China's Evolving Oil Diplomacy

China's oil diplomacy since the early 1990s has been marked by a new cooperative approach towards acquiring stakes in companies and oilfields overseas.[26] China has also been encouraging foreign companies to acquire stakes and participate in China's domestic and offshore energy fields, promoting cooperation and joint development of even the disputed energy fields. China's recent offshore oil exploration efforts have been centred around its Bohai Sea opposite Tianjin and the Pearl River Mouth area which are believed to hold over 1.5 billion barrels of oil reserves. Starting from March 2002, British Petroleum has already been developing Peng Lai drill in Block 11/05 from where commercial production is expected to reach 100,000 bbl/d. Similarly, Shell and CNOOC had been in negotiation over Bonan project in the Bohai Sea and CNOOC has already signed a production-sharing contract with Canadian Husky Oil for the Wenchang find in the Pearl River Mouth where production is expected to reach 500,000 bbl/d. All of this is aimed at building stakes with great powers in ensuring peace in China's immediate periphery.

Amongst other remarkable initiatives, after their boundary settlement agreement of December 2002, China and Vietnam have agreed to jointly develop Beibu Gulf (or the Gulf of Tonkin). China has indeed gone much further and CNPC has since acquired stakes in oilfields and production companies in Kazakhstan, Venezuela, Sudan, Iraq, Iran, Peru and Azerbaijan. The most significant deal of CNPC thus far has been its acquisition of a 60 per cent stake in the Kazakh oil firm, Aktobemunaigaz, with a pledge to invest more over the next twenty years. Similarly, CNPC has acquired stakes in companies like Greater Nile Petroleum Operating Company in Sudan from where oil imports have already begun to arrive as part of its investment deal. Russia is beginning to emerge as another major focus of Chinese oil diplomacy. Feasibility studies have been carried out to lay a pipeline between Anagarsk (in Russia) to Daqing with a capacity to carry 600,000 bbl/d of crude oil. Indeed, China now has been debating about creating its own strategic petroleum reserves of high capacity. The two US-led Gulf wars have really made this an issue of high priority for the Chinese leadership.

In addition to these cooperative strides, China has also tasked its naval forces to include the building of capabilities and strategies to escort tankers and to protect access to SLOCS and ensure safely on choke points, though they have also been equally focused on building their overall engagement with the Indian Ocean Region. In some of these Indian Ocean states, China is believed to have even developed (or obtained access to) some military facilities. If anything, this military-guided engagement with Indian Ocean rim states is not exclusively geared to the purpose of energy. Nevertheless, China's military engagement has emerged as a backbone for China's comprehensive engagement of Indian Ocean rim countries. Amongst the main "target" of China's military engagement one could include countries such as Burma, Bangladesh, Sri Lanka, Pakistan, the Maldives, Iran, Sudan, South Africa and Tanzania. In this context, China's Type 093 nuclear powered attack submarine – capable of firing land-attack cruise missiles with nuclear warheads and expected to be commissioned by end of 2004 or early 2005 – is going to add a new dimension of China's blue water navy with implications for Indian Ocean rim countries.[27]

Security Implications for the Indian Ocean

In addition to countries of the Indian Ocean rim, China has also tried to build its engagement with other major players in the global energy market that have been active in this region. In particular, China has focused on both inward as well as outward investments. China's State Owned Enterprises (SOE) have lately become aggressive in acquiring interests in oilfields and oil firms as well as in inviting FDI for the exploration of China's onshore and offshore gas and oil. Indeed, since developing offshore energy remains an expensive, high-tech and high-risk operation, this was traditionally the pioneering sector in China to open up to foreign enterprises. Nevertheless, this process still continues to move slowly, with several inbuilt limitations. In view of this growing interdependence, there remains a strong possibility of China's expanding energy deficit causing at least some benign threats as it continues to produce large quantities of unclean and polluting energy resources. China's widespread use of coal, for example, may cause an increase in acid rain and, such human-made disasters, of course, know no national boundaries. It is issues like this that may, in turn, become a major hurdle in China's FDI and the resultant economic stagnation may cause havoc for China's economic development leading to an outpouring of the Chinese population to neighbouring states and into other rapidly-developing Indian Ocean rim countries.

Threats can also be more real than some of these overstretched propositions. Chinese scholars no longer shy away from asserting that they do see a link between China's military ability and China's energy security imperatives.[28] Whether the Chinese Navy can help ensure the freedom of SLOCS and choke points has come to be critical to China's maritime thinking and planning.[29] They often cite how China could face an energy crisis if its oil supply lines at sea were attacked, especially in places like the Straits of Malacca where one oil-tanker blown up in a narrow section of the Straits can close the route for a long period of time, forcing China to opt for an expensive re-routing. This has since made SLOCs a priority agenda issue of China's security and foreign policy for the 21st century. This partly explains China's indulgence with littoral states in its immediate periphery moving from the South China Sea to the Indian Ocean rim.

Given the fact that four-fifths of China's imported oil comes through the Straits of Malacca, speculations about this choke point led to Chinese President Hu Jintao to propound his thesis of the "Malacca dilemma" which is seen as a key element in China's energy security vision.[30] President Hu's thesis is that the United States has all along encroached upon and tried to control all the navigation through the Malacca Straits. Flowing from there, Chinese experts have been expounding both defensive and offensive options for a new naval strategy: "One [option] is making quick reactions, including a military reaction, when a crisis occurs . . . to display the strength for safeguarding the country's interests. The other is the capability of reciprocal deterrence. This means if you can threaten my international shipping route, I can also threaten your security in various fields, including your international shipping route security."[31]

India's Security Concerns

While IOR-ARC as a body seems far more receptive to China's engagement in the Indian Ocean region, some countries such as India have reasons to feel concerned over China's forays around the Indian Ocean rim. Not only have China and India not had the best of relations in their recent past, but India is also witnessing an economic boom leading to a similar rise in its energy needs, thereby increasing its dependence on the import of energy which will become imperative for India's continued development in coming years. To appreciate how rising demand for clean energy may create friction between China and India, one has to see how consumption of natural gas in both China and India stood respectively at 0.5 Tfc and 0.4 Tfc 1990 and how it is now projected to reach 6.1 Tfc for China and only 3.4 Tfc for India by 2020.[32] This has clear implications for the nature of the options available to India to deal with any energy deficit crisis which may create newer flashpoints as China expands its access to Indian Ocean energy resources.

Given India's increasing import dependence on oil – which is expected to rise from the current 40 per cent to 70 per cent by 2010 and the import of gas is also likely to reach about 40 per cent of total consumption – it is very likely that rising demand in China, India, Japan, Taiwan, South Korea and so on will create a scramble for energy resources.[33] Like China, therefore, India is already finding itself under pressure to incorporate its military means as part of its efforts towards ensuring the safety of its energy supplies both in terms of dedicated suppliers or sources as well as supply lines, especially the SLOCS. Indeed, India has already been trying a similar twin strategy of building bridges with the Chinese through a series of confidence-building measures, while at the same time strengthening its naval capabilities to ensure its energy interests in the Indian Ocean rim. India's creation of a tri-service command at Fortress Andaman clearly indicates the way India is likely to evolve its response to some of these trends in coming years. At one level, therefore, the evolving Indian comprehensive approach may invoke a direct response from the Chinese side, or from other major energy-deficit states.

Already out of its total use of 2.0 million barrels per day, India's net oil imports account for about 1.2 million barrels per day. This consumption is projected to increase to 3.2 million barrels per day by 2010. Similarly, gas consumption has also moved from being 0.6 trillion cubic feet (Tfc) for 1995 to becoming 0.8 Tfc for 2000 and it is expected to rise to 1.8 Tfc by 2010 and by 2020 China is expected to reach 4.5 Tfc. It is certainly going to create pressures from both China and India that is bound to reflect in their energy policies and strategies both between themselves as well as vis-à-vis countries of the Indian Ocean rim. Apart from countries in India's immediate periphery where China's indulgence has always been an issue in India-China ties, their growing indulgence in Sudan can be seen as an example of how

China and India are driven by their energy needs to compete in the Indian Ocean rim.

From a more practical perspective, experts talk about energy and foreign trade being critical for both New Delhi and Beijing in terms of how they view the importance of the Indian Ocean to their respective national objectives. This is especially important for China, since all of its trade and energy supplies from Asia, Africa and Europe flow through the Indian Ocean SLOCS. Given India's rise in recent years, China has become particularly alive to protecting its access to these SLOCS even in face of a potentially hostile New Delhi.[34] Similarly, in view of China's likely competition with other major powers such as the US and Japan – and India's growing closeness with these two countries – Beijing sees its access to the Indian Ocean as part of its strategy to secure its backdoor exit that adds to the strategic depth of its naval forces. This explains China's indulgence with a wide range of Indian Ocean rim states where developmental projects and trade have strongly backed up military supplies, all aimed at fostering a client-benefactor relationship of dependency.[35]

Conclusion

To conclude, therefore, China's growing energy deficit remains an extremely complex reality and can unleash both cooperative as also confrontationist tendencies in China's foreign and security policies. Ideally speaking, given China's rising interdependence with the outside world, it must adopt a cooperative security strategy to obtain access to advanced technologies, energy supplies and to ensure safety of its supply lines. However, greater insecurity on the part of the Chinese elite is very likely to generate more confrontationist tendencies. Therefore, the future of China's energy deficit as well as its impact on the Indian Ocean region will depend on how China evolves its cooperative strategies vis-à-vis the Indian Ocean rim as well as how Indian Ocean rim states respond to China's initiatives. In addition, the policies and practices of other energy deficient states in the region will also have a major influence on the evolution of China's Indian Ocean policy.

In terms of its short-term strategy, China seems to have focused on building bridges with other major players in the energy scene in the Indian Ocean rim as well as to evolve alternative methods to exploit its traditional sectors such as coal in order to generate clean energy like electricity. This strategy is expected to take care of China's energy demands in the short-term, but beyond this, even its coal reserves are likely to become problematic as the costs of new technologies for making coal clean, efficient and environmentally-friendly will continue to rise. Similarly, the deficit may not make much difference to China's rising demand for gas (where China has abundant reserves) but its growing oil deficit carries the potential for unleashing energy related flashpoints in its periphery and especially in turmoil-ridden parts of the Indian Ocean rim. Furthermore, trends in China's own internal politics will also remain decisive.

While China's current energy strategies emphasise building bridges with major players in the energy markets abroad and exploiting alternative and renewable sources of energy at home, yet, by China's own estimates, with the rising global demand for energy and limited supplies, by about 2050, energy is likely to emerge as the most critical flashpoint in most inter-state ties. All of this has clearly strong security implications for the Indian Ocean region where energy consumption is likely to grow much faster and supply lines traversing so intensely in this region are bound to witness growing competition. The focus, therefore, has to remain on devising alternative and cooperative strategies for managing this competition as well as towards using available energy resources more wisely and in exploring alternative and renewable sources of energy which remain the only possible way out of this impending crisis.

NOTES

1. See for example, *Guojia nengyuan jieyue yu ziyuan zhong he liyong shiwu guihua* (The National Tenth Five Year Plan for Saving Energy and Comprehensive Use of Natural Resoruces), Guojia jin maowei, November 2001 (National Trade and Economic Commission, November 2001), at http://www.chinacp.com/newcn/chincp /policy-of-setc13.htm; *Guojia Fazhan jihua weiyuan hui shiwu guihua zhanlue yanjiu xiache* (National Development Plan Commission Strategic Study of the Tenth Five Year Plan), Beijing, Zhongguo renkou chubanshe (China Population Press, 2000), p. 531; Shang Weiguo, "Zhongguo shiyou: Shi guan zhan lue anquan", (China's Oil:

A Issue of Strategic/Security Importance), *Shijie Zhishi* (The World Knowledge), 2002, Issue 21, p. 34; Wang Jian, "Zhonguo nengyuan de qangqi gongqiu xingshi",(Long-term Situation of Chinese Energy), *Zhongguo hongguan jinji xingxi wang* (Information Network of China Academy of Macro Economics), 12 July 2001.

2. Erica Strechker Downs, *China's Quest for Energy Security*, RAND Paper MR-1244-AF, 2000, p. xi; World Bank, *China 2020: Development Challenges in the New Century*, (Washington DC, 1997), pp. 20-21; also for 2020 projections see http://www.phoenixtv.com/home/finance /fortune/200301/13/21660/html; compare these to earlier versions like those in Nigel Holloway, "For Whom the Bell Tolls", *Far Eastern Economic Review*, 2 February, 1995, pp.14-16; also "Business Briefing", *Far Eastern Economic Review*, 13 October, 1994, p.79.

3. Zhang Wenmu, "Zhongguo nengyuan anquan yu zhen che xuan zhe" (The Energy Security of China and the Policy Options), *Shijie Jinji yu Zhenzhi* (*World Economics and Politics*), Beijing, Issue 5 (May 2003), p.11.

4. Xia yi shan, "Dang qian guoji nengyuan xinshi he zhongguo nengyuan zhanlue", (Current International energy situation and Chinese Energy Strategy), *Heping Yufa Zhan* (Peace and Development Quarterly), 2002, Issue 2, p. 36.

5. Swaran Singh, *China-South Asia: Issues, Equations, Policies*, (New Delhi: Lancers Books, 2003), Chapter 4, 'China's Indian Ocean Policy', pp.100-120.

6. Keneth Lieberthal, *Governing China: From Revolution to Reform*, (New York: W.W. Norton, 1995), p.77; Kim Woodward, *The International Energy Relations of China*, (Stanford, CA: Stanford University Press, 1980), p. 33. No doubt China had been net importer of oil during 1950s from Soviet Union but the Sino-Soviet split and discover of Daqing oilfield had been reason behind this obsession with self-sufficiency.

7. In 1988 China reorganized most of its State owned oil and gas assets into two vertically integrated firms – the China National Petroleum Corporation (CNPC) and the China Petrochemical Corporation (Sinopec). Both were regionally focused and while CNPC controlled much of north and west, Sinopec controlled much of south. Similarly, while CNPC controlled much of crude oil production Sinopec controlled much of refining. Much of the offshore production – which has become the new focus – came under the China National Offshore Oil Corporation (CNOOC). Later, China National Star Petroleum was created later in 1997and State Energy Administration was created in 2003 and it is responsible for all the regulatory oversight.

8. Amy Myers Jaffe and Steven W. Lewis, "Beijing's Oil Diplomacy", *Survival* (London), Vol. 44 No. 1 (Spring 2002), p. 117.

9. For details on how issues can be securitized see Barry Buzan, *People, State and Fear*, (Boulder, Colorado: Lynee Rienner Publishers, 1991); Barry Buzan, Ole Waever, Jaap de Wielde, *Security: A New Framework for Analysis*, (London: Lynne Rienner Publishers, 1998).

10. Swaran Singh, "China's Changing Maritime Strategy: Implications for the Indian Ocean Region", *Journal of Indian Ocean Society*, (New Delhi), February 1998.

11. CNPCs high performing subsidiary – PetroChina – had carried out IPO during early 2000 through both Hong Kong and New York stock exchanges and successfully raised over $ 3 billion with British Petroleum holding 20 per cent of these. Since then several foreign players have been involved in China's oil and gas fields. Amongst those involved in Bohai and Pearl River Mouth offshore fields (which have implications for Indian Ocean) include Chevron Texao from 1999, Philips Petroleum from 2000, Shell and Husky Oil from 2001 while Shell and Exxon Mobile have also been involved in pipelines projects.

12. "Shell Launches a Major Investment Project in China", *The Peoples Daily*, 07 August 2002 at http://english.peopledaily.com.cn/200208/eng20020807_101086.shtml; also for regular official details on this see *China's Petroleum Industry* at http://www.vitrade.com/china/chinanews_brieing_oil_industry.htm. CNOOC was set up in 1982 and currently has developed 19 offshore gas and oilfields. Its current subordinate companies include Bohai Sea Petroleum Corporation, the South China Sea Petroleum Corporation, the West Petroleum Corporation, the East China Sea Petroleum Corporation, over a dozen professional contract companies, and manufacturing entities and research centers and institutions.

13. Projections for 2020 amongst others are from conservative ones like Shixian Gao, "China" in Paul B. Stares (ed.), *Rethinking Energy Security in East Asia* (Tokyo: Japan Center for International Exchange, 2000), pp. 43-58 or PRC State Council, *China Energy Strategy Study 2000-2050*, (Beijing, 2000) to highest projections, for example, from International Energy Agency, *China's Worldwide Quest for Energy Security*, (Paris: OECD0, 2000); And for figures for 1990-2000 see "Asian Demand Flat, and May Get Worse", *Petroleum Intelligence Weekly*, 30 July 2001, p. 2.

14. *Shijie jinji nianjian 2002-2003*, Beijing: Jinji kexue chuban she, 2002, (*World Economic Survey 2002-2003*, Beijing: Economic Academy Press, 2002), p. 400.

15. n.3, p. 14.

16. Even in India coal-based thermal power generation accounts for much less share at 70 per cent of total electricity generation in recent years. For further details see country study for "India" by Energy

Information Administration at their official website at http://www.eia.doe.gov/emeu/cabs/india.htm

17. n.15.

18. Zhang Weiping et al. (eds), *Twenty Years of China's Environmental Protection Administration*, (Beijing: China Environment Sciences Press, 1994), pp. 215-217; Barbara J. Sinkule And Leonard Ortolano, *Implementing Environmental Policy in China*, (Westport: Prager, 1995), p. 27; Abigail R. Jahiel, "The Organisation of Environmental Protection in China, *China Quarterly* (London), No. 156 (1998), p. 767.

19. China has decided to raise its budget for environmental protection from 0.93 per cent of GDP during the ninth five year plan to 1.2 per cent of GDP for tenth five year plan. See "New Five-Year Plan Gives Priority to Environmental Protection", *Renmin Ribao* (Peoples Daily) (Beijing), 13 March 2001, at http://english.peopledaily.com.cn/200103/eng20010313_64836.htm

20. "Hu: Mondernisation needs long uphill battle", *China Daily* (Beijing), 24 April 2004, p. 1. These figures do not include the foreign trade and forex reserves of Hong Kong which is now part of China. These also do not include similar figures for Taiwan which is potentially seen as part of China.

21. "Creeping Irredentism in the Spratly Islands", *Strategic Comments* (London:IISS, 1995), No. 3, 22 March 19995.

22. n.5, p. 109.

23. Ferenc A. Vali, *Politics of the Indian Ocean Region: The Balance of Power*, (New York: The Free Press, 1976), p. 193.

24. "ForMin Spokesman: PRC IORARC Entry To Boost Regional Ties", *Foreign Broadcast Information Service-China-2000-0125*, dated 25 January 2000.

25. Liu xinghua, Qing yi, "Zhongguo de shiyou anquan jiqi zhanlue xuanzhe" (The Oil Security of China and its Strategic Options), *Xiandai Guoji Guanxi* (Contemporary International Relations), Beijing Issue 12 (December 2002), p. 27.

26. Zhang Yirong, "China Quickens Its Pace of Overseas Operations", *China Oil & Gas*, Vol. 4, no. 3 (September 1997), p. 174; "David B. Ottaway and Dan Morgan, "China Pursues Ambitious Role in Oil Market", *Washington Post*, 26 December 1997, p. 1; "China Takes Control of Kazakhstan's Aktyubinsk", *East European Energy Report*, No. 69, (24 June 1997), p. 16; Ahmed Rashid and Trish Saywell, "Beijing Gusher, China Pays Hugely to Bag Energy Supplies Abroad", *Far Eastern Economic Review* (Hong Kong), 26 February 1998, p. 48.

27. Donald L. Berlin, "The Indian Ocean and the Second Nuclear Age", *World Policy Journal* (New York), 2003, pp.239.

28. n.15.

29. Liu Xinghua (and) Qing yi, "Zhongguo de shiyou anquan jiqi zhanlue xuanzhe" (The Oil Security of China and its Strategic Options), *Xiandai Guoji Guanxi* (*Contemporary International Relations*), Beijing, Issue 12 (December 2002), p. 27.

30. A story from PRC owned-newspaper *Wen Wei Po* (Hong Kong) was reproduced at http://www.afpc.org/crm/crm527.shtml

31. A story from PRC owned-newspaper *Wen Wei Po* (Hong Kong) was reproduced at http://www.afpc.org/crm/crm527.shtml

32. For details on statistics See *International Energy Outlook 2003* (Table A5) at http://www.eia.doe.gov/oiaf/ieo/tbl_5.html

33. G. Parthasarthy, "The quest for energy security", rediff.com at http://www.rediff.com/news/2001/jul/06gp.htm

34. John W. Garver, *Protracted Contest: Sino-Indian Rivalry in the Twentieth Century* (Seattle: University of Washington Press, 2001), pp. 275-312.

35. See for example, Happymon Jacob, "India-Sudan Energy Ties: Implications", *Observer Research Foundation*, http://www.oberverindia.com/analysis/AO31.htm

APPROACHES TO ENERGY INSECURITY

CHAPTER 10

Energy Cooperation in South Asia: Some Observations

A. Subramanyam Raju

Introduction

Energy security has been defined as the "availability of energy at all times in various forms, in sufficient quantities, and at affordable prices" (Andres-Speed et al, 2002, 13). It can also be considered "in terms of the physical availability of supplies to satisfy demand at a given price" (Naik et al, 2003). All aspects of these definitions are critical to an understanding of energy security and energy cooperation in South Asia.

This Chapter attempts to analyse the existing energy situation in South Asian states and explores the problems and prospects of regional cooperation in the energy sector and the resultant mutual benefits. In addition, the chapter discusses India's energy security scenario, the various gas pipeline proposals to India, and regional initiatives for cooperative efforts in the energy sector. In order to realise these aims, the chapter will first outline some of the background to South Asian energy security. It will then provide an overview of energy cooperation in South Asia. Third, the scenario for India's energy security will be considered along with India's overseas investment. Finally, there will be a discussion of bilateral energy cooperation in South Asia.

South Asian Energy Background

Energy security has become an important concern to most of the countries in the world. Energy availability is one of the foremost requirements for any state in present times in order to grow economically and to prosper. Thus, safeguarding energy security has become an important element in the life of states to meet the aspirations of their citizens. Energy is being considered as the hub of the next revolutionary technological change, transforming our lives (Kelkar, 2003). In this context, the importance of energy cooperation in South Asia has been increasingly recognized.

Economic reforms and population growth in South Asia are resulting in a rapid demand for more and more energy. The South Asian countries have low levels of energy production and per capita consumption and many people in the region still lack access to electricity for their basic needs. These countries are largely dependent on energy imports although they have natural resources in abundance that are yet to be tapped and exploited. Hence, at present, regional cooperation in the energy sector among these states is a desirable policy goal.

South Asia is important to the world energy market because it contains 1.3 billion people and it consumes about 5.9% of the world's commercial energy. However, this excludes all non-commercial energy sources – wood, animal waste and other biomass, which account for more than half of the region's total energy consumption. Regional energy needs will continue to increase with the ongoing liberalisation programmes in South Asia (Table 10.1).

Table 10.1. Projected requirement of fossil fuels.

Fuel	Units	1997	2010	2020
Coal	MMT	233	378	516
Oil	MMT	111	203	304
Gas	BCM	33	78	147

Source: World Economic Outlook, 2000

By 2020, South Asian coal consumption will be more than double the amount consumed in 1997 (516 million metric ton (MMT) compared with 233 MMT). In terms of oil consumption, it will be more three times the amount consumed in 1997 (304 MMT from 111 MMT), whereas gas consumption in 2020, will increase by more than four times 1990 consumption (147 billion cubic meters (BCM) from 33 BCM).

South Asia depends on coal (47%), petroleum (33%), Natural gas (12%), hydropower (7%) and nuclear power (1%) (Malla, 2004). There is a significant variation in energy mix by state in the South Asian region, however For instance, Bangladesh depends 69% on natural gas, India on coal by 55%, Sri Lanka and Maldives use 76% and 100% of petroleum respectively, Pakistan depends 43% and 38% on oil and natural gas respectively, and Nepal and Bhutan consume 90% of petroleum and hydropower (Malla, 2004).

In most of the rural areas of South Asia, people use non-commercial energy for their consumption. Although it causes air pollution leading to respiratory, heart and lung problems, it reduces the imports of oil and gas. However, this is nonetheless a serious problem of human health since it is estimated that more than half a million Indians die each year due to polluting traditional fuel. In India, the largest country in South Asia, there are 80,000 villages that are yet to be electrified. Out of these, 18,000 villages can be electrified by using renewable resources. Apart from the inability of the government to provide infrastructure, rural people are often not in a position to pay electricity bills and instead they use sources like firewood, cow dung, agricultural residue and kerosene.

Energy Cooperation in South Asia

On 7 August 1998 at the Dhaka Workshop on *Improving the Availability of Power in South Asia: Search for Optimal Technology Options*, the participants agreed that South Asia would prosper by sharing its energy resources (*Financial Express*, 1998). The workshop realised the importance of setting up a Regional Power Grid, which would ensure the quality of power supply in the region. South Asia could promote a South Asian Development Triangle, similar to others that have been developed in state peripheries in ASEAN. This would comprise the Northeastern parts of India, Bangladesh and Nepal; a Southern Grid would connect South India with Sri Lanka and a Western grid, especially including Punjab (India) would connect with Pakistan's Punjab.

Such a concept could develop into a growth quadrangle involving cooperation among Bangladesh, Bhutan, Nepal and the Northeastern parts of India. Since energy is one of the determinants of economic well-being in the new millennium, the South Asian states should take initiatives in sharing their energy resources. Though bilateral cooperation in South Asia is successful, there are many obstacles to forging multilateral cooperation.

Through multilateral cooperation, regional states would minimise the costs of research and improve efficiency. South Asia needs an integrated energy policy incorporating hydropower, oil, natural gas, nuclear and renewable sources of energy. For instance, there exists a proposal for a SAARC Power Grid to connect the national power grids of all the countries in South Asia. This would help to transmit power from one country to another country. Synergy among the neighbouring states should be developed to exploit the resources. The South Asian states should create an energy database and cooperate in a regional analysis of energy production, consumption, exports, prices and demand forecasting, which are required for the development of a regional energy market. It is also essential for them to establish energy codes, technical specifications and standards for all the countries in South Asia that would certainly soften the regional energy trade. It is also important to have an awareness programme in South Asia to educate political leaders, decision-makers and people regarding the benefits of regional energy cooperation and cross-border sales of electricity, oil and natural gas. The vision for a regional initiative for energy cooperation is to develop an energy market, maintain an effective and reliable energy supply, and thus pave the way towards sustained regional economic growth and poverty alleviation. To achieve this, all South Asian states should accept the joint development, trading and sharing of energy. Apart from Sri Lanka and Maldives, all other South Asian countries are geographically connected to one another. Thus, it is theoretically feasible to develop an integrated energy infrastructure including power grids and gas pipelines. There is a great deal of complementarity in the energy sector in South Asia. For instance, there is hydropower potential in India, Pakistan, Nepal and Bhutan, and abundant gas resources in Bangladesh and coal in India. This cannot only improve their efficiency but will also minimize oil imports, which will lead to a reduced dependency on foreign currency. Further, there are alternative energy resources – renewable resources such as solar, wind, and indigenous small hydro, biogas and biomass that could all play an important role in increasing access to electricity in rural areas. These countries should understand the benefits that could be made available through regional cooperation. To realise this, however, requires considerable mutual trust and cooperation among regional states.

At present, however, every country's approach towards energy planning and development is carried out from its own domestic point of view. As a result, all countries are overlooking the opportunities in their neighbourhood. Since India is the largest country in South Asia, it is relevant here to discuss its particular energy security scenario.

India's Energy Security Scenario

India consumes more energy than any other country in South Asia. It is the seventh largest consumer of oil in the world yet it produces only about 30% of its consumption. With the ongoing economic reforms in India the demand for energy consumption is likely to increase many times. To meet the future demand, the Indian government set up a group on *India's Hydrocarbon Vision 2025* in February 1999. The main recommendations put forth by the group were:

• 100% exploration coverage of Indian sedimentary basins by 2025
• Exploration in frontier and deep water areas.
• Offering of attractive fiscal terms.
• Overseas investment in exploration of oil and gas resources.
• Self-sufficiency in petroleum refining
• Storage of oil in various locations.

India's future energy requirements in all categories are substantial (Table 10.2). If we compare energy demand from 1990-2020, coal, oil, gas and hydropower are going to increase rapidly. For instance, based on these data, the demand for oil and coal will almost quadruple during the 1990-2020 period. India, in 2020, would require seven times more gas than what it consumed in

1990, and, in hydropower, it will require five times the amount it consumed in 1990. It is interesting to note that the demand for nuclear energy, crop residue, fuel wood and animal waste is not projected to increase by 2020. It is clear that India will have to depend on imported oil, gas and hydroelectric power from other countries.

Table 10.2. Energy supply projection for India.

Fuel	1990	2000	2010	2020
Coal:	3850	7145	10508	15516
Indigenous	3850	7145	10508	15167
Imports	--	--	--	349
Oil:	2416	4245	7553	10723
Indigenous	1410	1592	1774	1955
Imports	1006	2653	5759	8768
Gas:	480	1002	1909	3478
Indigenous	480	1000	667	533
Imports	--	2	1242	2945
Hydro	249	324	625	917
Nuclear	24	31	25	12
Total Commercial energy	7019	12747	20600	30646
Crop residue	763	763	763	763
Fuel wood	3134	3134	3134	3134
Animal waste	1314	1227	1113	939
Total traditional energy	5211	5124	5010	4836

Source: http:www.teriin.org/energy/security.htm

In 2002, India spent Rs. 84,000 crore to buy oil to meet 69% of its oil fuel needs (Jayaram and Dubey, 2003). Its crude oil requirement is seen as growing to 364 mtpa by 2025 from 100 mtpa at present. India's economy is growing especially as a result of the implementation of economic reform. If it has to grow at a rate of 8% as per projection there needs to be an uninterrupted supply of oil. As most of its oil comes from the Middle East, India is potentially vulnerable to any event that may cause instability in that region. Any disruption in the oil supplies from the region or any price increase by the oil producers of the region will likely have a negative impact upon India's economy.

India's Overseas Investment

India has realised the importance of investing in foreign oil and gas fields to secure its imports. The Oil & Natural Gas Corporation (ONGC), ONGC Videsh Lmtd (OVL), the Gas Authority of India Limited (GAIL) and private companies such as Reliance Company have entered into a production-sharing contact. Negotiations are going on for oil exploration in other countries. Table 10.3 shows India's investment in oil and gas reserves in different countries.

Having discussed the energy sector in India, in the following section an attempt is made to focus on the cooperation between India and the neighbouring South Asian states.

Bilateral Cooperation

Energy Cooperation between India and Bangladesh

Bangladesh contains natural gas reserves of around 16.3 trillion cubic feet (tcf). Further, it is estimated that there is an additional 32.1 tcf in undiscovered reserves. If these figures are proved to be correct, then Bangladesh could certainly become a major gas producer. Many people in Bangladesh want the gas resources to be used for domestic purposes and oppose exporting gas to other countries, particularly to India. There are two essential views on this matter. First, some people express the view that Bangladesh may have a substantial future domestic demand projection, and, second, some people fear that India may exercise hegemony over their country. Bangladesh is yet to decide to export gas to India.

Bangladesh will benefit if it exports gas to India. One analyst rightiy pointed out that "Dhaka's inclusive, and perhaps increasingly irrelevant debate on gas exports reflects the tragic sub-continental tradition of wallowing in isolation and poverty rather than creating shared wealth (*The Hindu*, 12 November 2002).

Energy cooperation between India and Pakistan

India's security relationship with Pakistan is very important since any gas pipeline to India from the Middle East would require Pakistan's involvement and partnership. We have seen a positive step in the recent Summit in Islamabad. Abdullah Yusuf, Secretary for Petroleum and Natural Resources, stated that "Pakistan may start buying up to 5 million tonnes of diesel from India instead of

Table 10.3. Indian companies' overseas investment.

Country	Companies	Estimated oil resources
Libya (Ghadames Basin)	OVL: 49% IPR (US): 51%	Estimated oil reserves of 645 million barrels
Sudan (Greater Nile)	OVL: 25% China National Petroleum: 40% Petronas: 30% Sudan National Oil Company: 5%	Estimated oil reserves of 1 billion barrels
Syria	OVL: 60% IPR (US): 40%	Reserves yet to be estimated
Iraq (Block 8 Western Desert)	OVL: 100%	Reserves yet to be estimated
Iran (Farsi Offshore)	OVL: 40% Indian Oil: 40% Oil India: 20%	Estimated oil reserves of 540 million barrels
Yemen	Reliance: 30% ONGC: 40% BP: 30%	Reserves yet to be estimated
Myanmar (A 1 block off Rakline Coast)	OVL: 20% GAIL: 10% Daewoo: 60% Kogas: 10%	Estimated reserves 14-16 tcf of gas
Vietnam (Nam Conson Basin)	OVL: 45% PetroVietnam: 5% BP: 50%	Estimated reserves of 2.04 tcf of gas
Russia (Sakhalin-I)	OVL: 20% Exxon-N: 30% Sodeco: 30% SMNG-S: 11.5% RN Astra: 8.5%	Estimated reserves of 1.15 billion tones of oil and 20 tcf of gas.

Source: Anup Jayaram & Rajeev Dubey, "India's Oil Hunt", *Business World*, Vol. 23, No. 7, 8-14 July 2003, pp. 28-29.

Kuwait because it makes "better economic sense" – that is, it would prove to be cheaper (Jha, 2004). Yaswanth Sinha, India's External Affairs Minister, in an address to the SAARC Chambers of Commerce and Industry in Islamabad on 2 January 2004, expressed the view that "If Pakistan can find within itself the strength and wisdom to change its current approach towards India, .there are immense benefits that it can derive as a transit route for the movement of energy, goods and people" (*The Hindu,* 3 January 2004, 11). If this became a reality then both countries will obtain mutual benefits. In addition, increasing economic interdependence will make it easier for the two countries to resolve their political problems.

Pakistan has a 500 kV primary transmission system extending from Jumshoro in the south to Tarbela and Peshawar in the north. All of the lines are close to India's borders and may not require very complex transmission extensions. Pakistan's Power Minister stated that "There is a complete network on our side, of course, on their (India) side as well. What we need are the connections, which would take a couple of weeks" (*Hindustan Times,* 1999). If Pakistan exports its surplus of 200 MW, it will receive US$1.2 billion a year over a period of 20 years (Lama, 2000). Furthermore, it can help the power plants to use their power at 80% capacity compared with the present 60%. If the plants run at 80% capacity utilisation, the tariff would decrease to 3.5-4 cents. However, Pakistan is demanding Rs.3.00 (7 cents) per unit, whereas India is willing to pay for Rs.1.26 (3 cents) per unit. Pakistan wants to sell power in cents and not in rupees. This will require the setting up of foreign exchange reserves. India receives Rs.1.30-1.50 per unit from the National Thermal Power Corporation (NTPC), whereas the Pakistan government is paying Rs.3.60 per unit to the Independent Power Produces (IPPs). It has to balance both the internal and external prices. India cannot buy power if it goes beyond Rs.1.50 per unit. There has been an asymmetry in the power tariff between these two countries. Both states need to sort out their differences and, as a result, both will benefit from a viable power deal. If Pakistan exports power to India, it will result in a major transformation in the political economy of South Asia.

The India-Pakistan-Iran Pipeline

India and Pakistan signed a Memorandum of Understanding in July 1993 regarding exploring the prospects of a natural gas pipeline from Iran to India via Pakistan. In 1995, Pakistan and Iran signed an agreement to construct a pipeline from South Pars gas field in Iran to Karachi in Pakistan with an extension to India. However, due to various factors it has not materialised. Pakistan insists on an onshore gas pipeline, whereas India prefers an offshore pipeline project.

It was estimated that a 2700 km overland gas pipeline through Pakistan would cost about $3.1 billion. Out of this amount, India would have to bear about $1 billion and the rest would be met by Pakistan and Iran. If the project materialised then Pakistan will obtain $700 million as transit charges from India every year (Pathasarathy, 2001). India has reservations about the onshore pipeline project on the following grounds:

a) India is keen on full-fledged economic relations with Pakistan. This would be possible only if Pakistan gives MFN status to India. India has already granted MFN status to Pakistan in 1995.
b) India has doubts about the safety and security of the gas pipeline in view of the ongoing tensions between the two countries.
c) India has to live at the mercy of Pakistan
d) There may be threats from Pakistan fundamentalists and terrorists to the pipeline.

In spite of these reservations on the part of India, the proposed project can still be pursued on the grounds that it is mutually beneficial for both states. Even the fundamentalists and disruptive elements may not obstruct the proposed line if it promotes the good of the people and reduces the problems of suffering.

India is keen to have an offshore pipeline to bring gas from Iran to the Indian west coast and is ready to construct a 2900 km offshore pipeline. In such a case, India does not need to pay any transit charges to Pakistan. Though initially it would require more investment, in the long-run the cost would be marginalised. It will require the definition of the continental shelf of Iran, Pakistan and India. However, Iran and Pakistan are keen on the onshore option, which is considered to be cheaper. The best option for India is to import gas through pipelines, because it reduces by 35% the cost compared to Liquefied Natural Gas (LNG) import through sea tankers (Prakash, 2004).

Energy Cooperation between India and Sri Lanka

Sri Lanka has been facing a power shortage. Though surrounded by water, it does not have any hydro potential to be exploited. Thus, Sri Lanka is looking for alternative sources to meet its demands. India and Sri Lanka signed a free trade agreement in 1998 designed to promote mutual trade. The former Sri Lankan Prime Minister, Ranil Wickremesinghe, while delivering a lecture in Chennai in August 2003, called for the development of the South India-Sri Lanka sub-region as a single market, which would promote economic growth in both countries (*The Hindu*, 24 August, 2003, 8).

Sri Lanka proposed a bridge between Rameswaram in Tamil Nadu and Talaimannar in Sri Lanka which would create road and rail links. The bridge could "carry transmission lines to hook up Sri Lanka to India's Southern Region Electricity Grid with the Kudankulam nuclear power plant serving as a base load station (Verghese, 2003, 10). Both countries can cooperate in the area of power if an electric grid stretching from Nepal to Sri Lanka is established (Wickremesinghe, 2003).

There are about 60 Indian companies that have invested five to six billion Sri Lankan rupees in sectors such as natural gas and oil, shipping, education, banking, textiles, and in the financial and service sectors in Sri Lanka (The Hindu, 10 August 2003, 10). It is reported that the firm NEXANT carried out a study and examined a 30 Km submerged High-Voltage Direct Current (HVDC) link through Palk Straits between the two countries for the export of 500 MW power to Sri Lanka. The Ceylon Petroleum Corporation gave 100 retail outlets to Lanka IOC, a subsidiary of Indian Oil Corporation Limited. Further, it is ready to give 150 dealers franchise retail outlets (Naik, Sajal and Raghuraman, 2003).

Energy Cooperation between India and Bhutan

India and Bhutan had an agreement related to hydroelectric power supply. Under this agreement Bhutan would supply surplus power to India from its Chukha Hydro Electric Project

(336 MW). There is another project – the Tala Hydro Electric Project (1020 MW), which is under construction and the surplus power from this project will also be supplied to India. Bunakha (180 MW) and Wang Chu (900 MW) projects are planned and the Sankosh project (4060 MW) is under consideration. Bhutan exports 80 per cent of its energy generation to India, which accounts for 30 per cent of its GDP. Through the Chukha Project, Bhutan earned Nu 1,454 million as profit in 1998-1999 from its power exports to India. As a result, it is in a position to increase revenues through the supply of power to India and to its own industries in the country. India provided technical and financial to this project. Its share was Rs.2.45 billion with a 60% grant and a 40% loan. Bhutan was able to redeem the entire cost of the project in five years. In return, India is able to supply power to West Bengal, Orissa and the Northeastern parts of India. However, India needs to upgrade its transmission lines to increase its access to Bhutanese power.

Energy Cooperation between India and Nepal

Nepal possesses the largest hydropower potential (83,290 MW) in South Asia. However, due to various reasons, Nepal is not able to tap its hydro resources and is able to install only a 283 MW capacity. There are proposals between India and Nepal for hydroelectric cooperation. Projects like the Pancheswar Project, which has a capacity of about 6000 MW, needs special institutional and financial assistance. The other project – West Sati (750 MW), is also proposed to develop to export power to India. Due to factors such as technical, production and supply constraints and political differences, cooperation between the two countries has not yet been reached.

Conclusion

The South Asian countries can create a better future for their people through mutual understanding and cooperation. They can address the common problems being faced by them by sharing their experiences, by undertaking collaborative analysis of the current situation and by assessing the future scenarios and prospects for energy needs. Energy security is as much a collective effort as military security. A South Asian Forum on Energy (SAFE) needs to be established to reduce the cost of research to improve energy efficiency and to promote energy cooperation among the countries of the region. This in return will benefit them and the relations will be improved in a win-win situation. While India can benefit from electricity and gas imports from neighbouring countries to foster its development, Bangladesh, Nepal and Bhutan can also accrue economic benefits from the export of hydroelectric power and natural gas. Pakistan's economy cal also benefit from its export of electricity to India. Through mutual help and cooperation, all regional states can reduce their dependence on energy imports. This in return will help in improving their political relationships. These states can learn a great deal from the experience of the European Union. The genius of Jean Monnet, the father of European integration, focused on key strategic resources, which had traditionally fuelled wars in Europe and later became the basis for integration. Of course, the European experience may not be applicable to South Asia in its totality, as they have to go a long way in pooling their strategic resources. At the recently held SAARC summit at Islamabad in January 2004, the South Asian countries signed the South Asian Free Trade Agreement (SAFTA) creating a Free Trade Area in the region. This is a forward step in the right direction for a cohesive, mutually beneficial relationship and may lead to the transformation of economic relations in the region. Civil societies of the region can also play a positive role in forging stronger energy cooperation.

References

Andres-Speed Philip, Xuanli Liao and Roland Dannreuther (2002), *The Strategic Implications of China's Energy Needs*, New York: Oxford University Press.

Financial Express (1998), 8 August.

Jayaram Anup and Rajeev Dubey (2003), "India's Oil Hunt", *Business World*, Vol.23, No.7, 8-14 July.

Jha Prem Shankar (2004), "Big Gulps of the Peace Pipe", *Outlook*, Vol.XLIV, No.3, 26 January.

Kelkar Vijay L. (2003), "Long-Term Energy Security for India", *University News*, Vol.41, No.33, 18-24 August.

Lama Mahendra P. (2000), "Economic Reforms and the Energy Sector in South Asia: Scope for Cross-Border Power Trade", *South Asian Survey*, Vol.7. No.1, January-June.

Malla Shanker K. (2004), "Energy Cooperation in South Asia", see http://www.cpd-bangladesh.org/saceps/ energy.

Naik Anant V, Sajal Ghosh and V.Raghuraman (2003), *Energy Security Issues and India*, July, see www.acus.org/energy.

Pathasarathy G. (2001), "The Quest for Energy Security", see http://www.rediff.com/news/2001/jul/106gp.htm.

Prakash Om (2004), "The Oman-Indian Gas Pipeline Project: Need to Reassurect Again", see ipcs.o.../print Military.jsp?action=show views+K

The Hindu, 12 November 2002 and 24 August 2003.

The Hindustan Times, 16 January 1999.

Verghese B. G. (2003), "For an Ocean Outlook", *The Hindu*, 25 November.

Wickremesinghe Ranil (2003), "Making Our People Rich", *The Hindu*, 25 August.

CHAPTER 11

The Energy Challenge in Small Island States and Territories: The Case of the South-West Indian Ocean Small Islands

Christian Bouchard

Introduction

Small island developing states (SIDS) and comparable small insular territories are of limited sizes, possess vulnerable economies, rely on a limited local resource base and are environmentally fragile.[1] Among several other constraints to their development is that, while they usually do not possess fossil fuel resources, they nonetheless rely principally on them to fulfill their energy needs.[2] This is exactly what is found in the small island states and territories of the South-West Indian Ocean where fossil fuels account for some three quarters of the combined total primary energy consumption of Comoros, Mauritius, Mayotte (France), Réunion (France) and Seychelles. Even in Réunion and Mauritius, where energy production from local sources is the most developed, the islands must import more than three quarters of their total energy needs (in terms of primary energy requirements).

However, the potential for renewable energy has yet to be tapped and, considering the high cost of the fossil fuels solution, the prospects for renewable energy sources appear to be appealing. If the dream of 100% renewable energy in the islands is out of reach for now, major progress can certainly be made to increase local energy production, develop energy efficiency and realize energy savings. This means that it is very possible to reduce the islands energy deficit as well as their energy imports. It is on this premise that we decided to produce this first-ever regional assessment of the energy situation in the small islands of the South-West Indian Ocean. After the presentation of the regional energy balance, the analysis of the needs, the local energy production and the renewable energy potential, we propose a scenario (the energy consumption transition) that leads to the progressive return of energy self-sufficiency in the studied islands.

Finally, one may note that these remote and small islands are part of those territories that do not possess fossil fuels but are heavily dependent on them. However, what is different from most of the other communities in this situation is that they are isolated and thus have to develop their own energy infrastructure and electricity generation. In this regard, the islands can be considered as closed systems. Thus, local solutions for energy supply are certainly much more important there than on continents where community energy supply is generally associated with a larger national or regional framework. This may well give the small islands an opportunity to progress faster than the rest of the world in the field of renewable energy and in progressing towards energy self-sufficiency.

The Small Islands of the South-West Indian Ocean

The study area includes the small island developing states (SIDS) of Comoros, Mauritius and Seychelles as well as the French Overseas Department of Réunion and Departmental Collectivity of Mayotte (Figure 11.1, Tables 11.1 and 11.2). Even if they have their own individual aspects and specific context, these islands share, in varying degrees, the development constraints and vulnerabilities typical of small islands. Apart from their small size, they are all quite mountainous, and, with the exception of the Seychelles (granitic islands, coral cays and atolls, with equatorial climate for the main islands), they are all volcanic in nature and subject to a tropical climate (warm all year round, trade winds and tropical cyclones). Thus, due to their respective geology, none of the islands possesses fossil fuel resources; but some may develop geothermal resources. Nevertheless, because of their inter-tropical position, island context and relief, the potential for hydro, solar, wind and ocean energies is quite high.

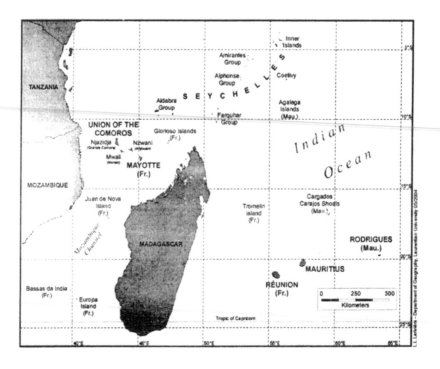

Figure 11.1. The small islands of the South-West Indian Ocean.

Table 11.1. Basic physical data for the small islands of the South-West Indian Ocean.

	Land area (sq km)	Islands	Geology	Highest point (m)	Coast-line (km)	EEZ Area (sq km)
Comoros	2,170	Njazidja (Grande Comore), Mwali (Mohéli)	Volcanic	2,360	340	161,993
Mauritius	2,040	Mauritius, Rodrigues (b), Agalega (c) and Saint- Brandon	Volcanic	828	177	1,274,638
Mayotte	374	Grande-Terre and Petite-Terre	Volcanic	660	185	50,000
Réunion	2,512	One massive island	Volcanic	3,069	207	309,956
Seychelles	455	115 islands	Granitic or coral islands	905	491	1,288,643

(a) The Union of the Comoros also claims French-administered Mayotte. (b) Rodrigues Island covers 119 sq km and its population is of 35,000 inhabitants. (c) Two islands covering 70 sq km in all and with a permanent population of some 300 peoples. (d) Dependency with no permanent population (Mauritius also claims the UK-administered British Indian Ocean Territory and the French-administered Tromelin Island). (e) Granitic islands for the main group (Inner Islands), coral islands otherwise (Outside Islands). Only the main group is inhabited by permanent communities.

Main sources: CIA, *The World Factbook 2004* (land area, highest point, coastline); WRI, Earth trends: the environmental portal (claimed exclusive economic zone in 2000, referring to the Global Maritime Boundaries Database -GMBD); Godard (1998), Atlas de France, vol. 13 : *Les outre-mers*.

In terms of the human environment, they are all quite densely inhabited although there are two different demographic and socio-economic profiles. On the one hand, the islands of the Comoros Archipelago (Njazidja, Mwali and Nzwani of the Union of the Comoros and the French Departmental Collectivity of Mayotte) have a very young and rapidly-growing population, as well as an underdeveloped economy and a low level of human development. On the other hand, Mauritius, Réunion and Seychelles have an older and slower-growing population as well as a more modern and diversified economy and a medium to high level of human development.[3] Regarding culture and ethnicity, the Comoros Archipelago is part of the Swahili world (inhabited by a Muslim indigenous population) while the Seychelles and the Mascarene Archipelago (Mauritius, Rodrigues and Réunion) are Creole islands (inhabited by a multiethnic population of European, African, Malagasy, Indian and Chinese origin with a diverse degree of interbreeding).

Table 11.2. Basic human data for the small islands of the South-West Indian Ocean.

	Population (mid-2004)	Density (inh./sq km)	Growth rate (%)	Median age	GDP 2002[a]	GDP per capita[a]	HDI (2002)
Comoros	652,000	300	3.5	18.6	1.0	1,690	0.530
Mauritius	1,220,000	598	1.0	30.1	13.1	10,810	0.785
Mayotte	186,000	497	3.2	16.9	[0.4][b]	[2,500][b]	[0.5-0.7]
Réunion	766,000	304	1.4	26.5	[10.7][b]	[14,800][b]	[0.8-0.9]
Seychelles	81,000	178	1.0	27.3	1.4[c]	17,030	0.853

(a) Gross domestic product in billion of purchasing power parity US$ (PPP US$); Gross domestic product per capita in purchasing power parity US$ (PPP US$). (b) Estimations based on available data (INSEE, CIA). (c) Data for 2001.
Main sources: CIA, *The World Factbook 2004* (population, population median age); PRB, 2004 World Population data Sheet (population growth rate); UNDP, *Human Development Report 2004* (GDP and GDP per capita in PPP US$, Human development index - HDI).

Energy Balance and Energy Dependency

Collectively, the five studied small island states and territories used some 2,500 ktoe (ktoe: one thousand tonnes of oil equivalent[4]) in primary energy in 2000 (Table 11.3). This figure is calculated by adding up the well-known energy balances of Mauritius and Réunion and the estimated energy balances for the Comoros, Seychelles and Mayotte. In the latter two islands, the estimations are based on the known oil product consumption and can be considered to be very reliable as the local energy production is limited. In the case of the Comoros, the data are a rough estimate because local energy production is not well-known even if it is still dominant in the national energy balance.

As the combined local energy production amounted to some 600 ktoe in 2000, the coefficient of energy self-sufficiency for this small island community is only 24%. At the same time, the coefficient of energy dependence reaches 76% when 1,900 ktoe of hydrocarbons were imported in 2000. These imports have been dominated by diesel oil (23-25% of the regional imports), coal that is only used in Mauritius and Réunion (18-20%), jet fuel (17-18%) and gasoline (14-15%), but also included are other fuels notably used by the energy sector for the thermal electric stations (24-26%) as well as some kerosene not used for aviation, and liquefied petroleum gas (especially butane and propane).

In terms of national balances, all of the islands, excluding the Comoros, are largely reliant on imports to fulfill their needs. Even in Mauritius and Réunion, where the local energy production is considerable (respectively 277 ktoe and of 174 ktoe in 2000), both islands have to import at least three quarters of their total primary energy needs. This is taking into account the energy used on the islands as well as the energy needed for aviation and maritime activities.[5] The situation is quite different in the Comoros where, in a context of poor electricity production and limited transportation (both local and international), fuel wood and charcoal still contribute approximately three quarters of the national primary energy supply (UDC, 2002, p. 2). As for Mayotte and Rodrigues, they are in an intermediate situation in which electricity is replacing fuel wood and charcoal as the households' primary source of energy, and transportation is developing rapidly. Finally, in Seychelles, local energy production only accounts for less than 5% of the island's total requirements.

Regarding Mauritius and Réunion, some key figures that are of interest for the last decade show that local energy production has not really progressed over the same period while the total needs have increased respectively by 50.7% and 64.6%. In Mauritius, 292 ktoe was produced from local energy sources in 1993 and 282 ktoe in 2003, for a ten-year mean of 283 ktoe. In Réunion, 248 ktoe was produced from local energy sources in 1982, 253 ktoe in 1990 and about 250 ktoe in 2000 (when, for that last year, converting the local hydro and solar electricity generation with the conversion factor used before 2000[6]). As the energy production from local sources remained quite stable over the period, thus energy imports rose from 530 ktoe in 1993 to 899 ktoe in 2002 in Mauritius (an increase of 70%) and from 408 ktoe in 1990 to 838 in 2000 ktoe in Réunion (an increase of 105%). This means that their coefficient of energy dependence rose respectively from 65% and 70% to the actual 75.4% and 82.8%; a deterioration of the coefficient of some 10 percent for both islands over the last decade.

Table 11.3 Energy balance in the small islands of the
South-West Indian Ocean, 2000.

	Comoros (a)	Mauritius (b)	Mayotte (c)	Réunion (d)	Seychelles (e)
Total primary energy requirements	168	1,126	70	1,012	105
Local energy production (ktoe)	131	277	< 10	174.4	< 5
Energy imports (ktoe)	37	849	60	837.5	100
Coefficient of energy self-sufficiency (%)	78	26.4	< 15	17.2	< 5
Coefficient of energy dependence	22	75.4	> 85	82.8	> 95

Estimation made on the basis of the national oil products imports (data from EIA, Internet) and the DGE (2002) assessment of the shares of local energy production and energy imports at the national level. (b) Including Rodrigues Island that accounts for about 1% of the national primary energy requirements (CSO, 2003, Table 2.1). (c) Taking into accounts the 60,809 cubic meters of oil products used that year, excluding army needs (INSEE Mayotte, 2003, p. 117). (d) Including some 187.3 ktoe for the

178,741 tonnes of jet fuel used (INSEE Réunion, 2003, section 16). (e) According to Juliette (2004), Seychelles consumption of final energy was of 102,148 toe in 2002 while the final consumption was of 72,747 toe, these two numbers only taking into accounts the oil products.

The same pattern of relying almost uniquely on fossil fuels to cope with the steady increase in energy needs is clearly evident in Seychelles, Comoros and Mayotte, while the development of wind generated energy is now underway in Rodrigues. As shown by the Mauritius and Réunion cases in the last decade, the fossil fuel solution will result in a significant increase in energy imports, the energy deficit and the coefficient of energy dependence in the coming years in Seychelles, Comoros and Mayotte. Taking into account tourism development and resultant increasing air traffic, this means that the energy deficit is also likely to continue to grow in Rodrigues, even with the commissioning in 2004 of the new wind farm.

Final Energy Consumption

In the studied small islands, energy consumption is characterised by a rapid growth that is related to the combination of population growth, economic development and the increasing ratio of energy consumption per inhabitant. For example, the final energy consumption has grown from 650 to 765 ktoe in Mauritius from 1993 to 2002, representing an increase of 17.7%, and from 640 to 967 ktoe in Réunion from 1990 to 1998 (data using the old conversion factor), representing an increase of 51.1% (Table 11.4). In Réunion, when adding jet fuel to the official figures, the final energy consumption reaches some 798 ktoe in 2000 (data using the new conversion factor). In this island, while population growth from 1990 to 1998 was 16.5% (596,000 to 695,200 inhabitants), the household energy consumption increased by 59% (127.8 to 202.7 ktoe). In addition, the economic sector (agriculture, industry and services, excluding transportation) consumption increased 30% (239.5 to 310.5 ktoe) and the transportation energy consumption increased 66% (272.8 to 453.8 ktoe).

At the regional scale, the quantity of final energy consumed and the consumption growth are strongly influenced by transport energy requirements. As shown in Table 11.4, the transport sector accounts for some 60% and 48% of the final energy consumption in Réunion and Mauritius respectively. In Mauritius, the transport sector consumed 256 ktoe in 1993 and 364 ktoe in 2002, an increase of 42% over the decade while the island's total final energy consumption only increased by 18% in the same period. In Réunion, the transport sector consumed 273 ktoe in 1990 and 454 ktoe in 1998, an increase of 66%, while the island's total final energy consumption only increased by 51% in the same period.

Table 11.4. Final energy consumption by sector, Réunion and Mauritius (2000).

	Réunion ktoe	Réunion %	Mauritius ktoe	Mauritius %
Household	117.6	14.7	99.2	13.2
Agriculture	0.7	0.1	4.8	0.6
Industry	93.7	11.7	249.9	33.4
Services	99	12.4	36.9	4.9
Transport	487 [a]	61.0	355.9	47.5
Other (not specified) and losses	–	–	2.3	0.3
Total final energy consumption	798	100	749	100

(a) Including an estimated 187 ktoe of jet fuel.

In Seychelles, transportation requires about the same quantity of energy as the power industry that used 45% of national primary energy requirements in 2002 (Juliette, 2004, p. 24). In terms of final energy consumption, jet fuel (jet kerosene) alone accounts for 40% of the archipelago needs, gasoline for a small 3% and diesel used by boats for approximately 20%.[7] Elsewhere, transport energy needs are booming in Mayotte as the island entered into a phase of very rapid economic development, and this situation should continue for at least the next few decades (as the island progresses towards stronger integration into the French Republic). During the same period, transport energy needs should also increase quite rapidly in Rodrigues as the island relies on tourism to induce economic development, and as well as in the Comoros islands as they recover from the serious global crisis that has paralyzed them since 1997.

Overall in the islands, the main problem with the transport sector is that it actually uses only fossil fuels, whether it is gasoline, diesel or liquefied petroleum gas (LPG) in domestic transportation, jet fuel in aviation or bunker fuel in maritime activities. However, fossil fuels are not only used as final energy in transport. For example, fuel oil (59.0 ktoe), diesel (37.8 ktoe), LPG (3.8 ktoe) and coal (16.1 ktoe) combined accounted for 47% of the final energy consumed by the manufacturing sector in Mauritius in 2002 (CSO, 2003). Kerosene (8.8 ktoe) and LPG (42.1 ktoe) contributed to 51% of household needs, while LPG (4.9) represented 12% of the service sector requirements, and diesel (2.5 ktoe) accounted for 51% of agricultural final energy needs. Altogether, these other demands for fossil fuels, excluding the transport sector, amounted to 175 ktoe, which represents some 29 ktoe more than all of the electricity consumed in the country (146 ktoe). The fossil fuels used for other needs (excluding the transport sector) are of lesser importance in Réunion where they amounted to 86 ktoe in 2000 (INSEE Réunion, 2003).

Local Energy Production

In the small islands of the South-West Indian Ocean, local energy production never contributes more than one quarter of the primary energy requirements, except in the Comoros islands, where it represents some three quarters of total needs (in a context of general crisis and energy shortage). At the regional level, local production amounts to roughly 600 ktoe, representing some 24% of the global primary energy requirements. In terms of sources, local generation breaks down approximately as follows: 54-56% for bagasse[8] which is now valorized in Réunion and Mauritius, 28-32% for wood and charcoal which are especially significant in the Comoros archipelago, 11-12% for hydroelectricity, essentially produced in Réunion and Mauritius, and 2-3% for solar (photovoltaic and thermal). In 2000, wind power, ocean energy, geothermal potential and domestic wastes were not used to produce electricity. While there is still a possibility to increase electricity production from bagasse by maybe as much as 50% in Mauritius, there is still an interesting potential for small hydro exploitation in Réunion, while hydro power development is now considered to be a priority in the Comoros. If wood can be considered as fully exploited in all of the islands, the situation is particularly difficult in the Comoros where the harvesting of wood has lately become more important than its renewal. Even if recent electrification has reduced the need for wood and charcoal in Mayotte and Rodrigues, these islands, like the Comoros, are now suffering from large-scale deforestation.

The under-exploitation of renewable energy is obvious in all of the studied islands. In the last two decades, the increase in energy needs has been almost uniquely met by the importation of more fossil fuels. It could not have really been otherwise for the transport sector, as no adapted alternative was proposed. In addition, the principle that new renewable energy on small islands is a feasible option in terms of technology, economy, environment and organisation has just been demonstrated. For now, only hydropower has been widely developed in the islands where it has been possible. The net result is that fossil fuels are still quite important in the manufacturing sector (for example in Mauritius) and represent the main, if not the unique, solution to electricity production.

Even in the bagasse and hydro-rich Réunion, only half the commercial electricity generation came from local energy sources in 2000, while burning coal and fuel oil produced the remaining 50% (Table 11.5). Elsewhere, fossil fuels accounted for 72.4% in Mauritius in 2002, more than 90% in the Comoros, and 100% in Seychelles, Mayotte and Rodrigues. In parallel, neither of the photovoltaic or the solar thermal technologies has yet been seriously promoted and encouraged at the household level other than in Réunion, even if they could significantly lower the global household sector requirements in commercial electricity. On this issue, Réunion is certainly an exception in the region with more than 40% of its households already equipped with solar domestic hot water systems, solar heating systems in several individual dwellings and a wide variety of photovoltaic uses.[9] New technologies for solar cooling systems are now also under experimentation.

Table 11.5. Commercial electricity generation, Réunion (2000) and Mauritius (2002).

	Réunion GWh[a]	Réunion %[a]	Mauritius GWh[b]	Mauritius %[b]
Hydroelectricity and photovoltaic	562	32	85.9	4.4
Bagasse	316	18	452.1	23.2
Coal	457	26	505.5	25.9
Fuel oil, diesel, kerosene	422	24	905.4	46.5
Total electricity available for sales	1,757	100	1,948.9	100

(a) Calculated production for each source from the total island production (CPI, Internet) and their respective percentage (Vantal, 2004).
(b) Including the 22.6 GWh generated in Rodrigues, which represented 1.2% of the Mauritian electricity production (CSO, 2004).

One of the worst adverse effects of the fossil fuel solution to electricity generation is the large quantity of energy lost in the energy conversion process. This is the main factor explaining the huge differences between the final energy consumption and the primary energy requirements (Table 11.6). In terms of global economic performance and development, the fossil fuel solution implies a large energy dependency that definitively represents a risk for the small island systems. It makes them very vulnerable to market price increases as well as to any disruption to supply. Relying on fossil fuels to cope with energy requirements is environmentally costly, especially in the context where the islands should reduce and not increase their greenhouse gas emissions. This is a very important issue as they claim that they will be severely impacted upon by the forecasted global environment change (especially in terms of projected sea level change but also by the strengthening of cyclones and possible changes in their wind and precipitation regimes). As they also requested international support to help them to reduce their vulnerabilities and be better prepared to adapt and successfully cope with the changes, they certainly have no other choice but to do their part and even be leaders in the international community in the field of greenhouse gas emission reduction.

Table 11.6. Primary energy requirements and final energy consumption, Réunion and Mauritius (2000).

	Réunion [a]	Mauritius
Primary energy requirements (ktoe)	1,011.9	1,125.9
Final energy consumption (ktoe)	798.3	749.0
Energy sector consumption and losses (ktoe)	213.6	376.9
Ration of energy efficiency: final energy consumption expressed in % of the primary energy	78.9	66.5

(a) Including, for both the primary energy requirements and the final energy consumption, some 187.3 ktoe of jet fuel..

If strong rhetoric is already there for a radical change in the way the islands deal with their energy supply, the conversion to renewable energy is slow to instigate. For example, in Seychelles, where all electricity is generated by gasoil combustion, the Electricity Division of the Public Utilities Corporation (PUC) has recently invested US$48.7 million in a new 50 MW thermal power station built on Mahé Island. This investment doubled the generating capacity of the country and also increased the capacity on Praslin and La Digue through the transfer of the existing smaller generator sets from Mahé to Praslin's power plant (Virtual Seychelles, 2003). In Mayotte, where all of the electricity is also generated by gasoil combustion, the thermal power plant capacity grew from 11.17 MW in 1995 to 39.6 MW in 2002 (INSEE Mayotte,

2003, chap. 17.2). In Réunion, the development of the co-generation bagasse-coal power plants in the 1990s allowed them to utilise the bagasse energy potential, but now that this resource is fully exploited, the actual power generation growth is reliant on an increase in coal combustion.

The Option of Renewable Energy

The goal of 100% electricity generation from renewable energy sources could certainly be met more easily than that of complete energy self-sufficiency and would already be a great achievement. The high price for the fossil fuel solution combined with the limited power demand in the small islands increases the unit cost of production for conventional thermal electricity generation. This creates a competitive situation for renewable energy technologies, especially in the context where technical solutions have now been developed for the specific maritime and tropical environment of the small islands.

For example, major technological breakthroughs have been made by the development of low and medium-powered wind turbines that can withstand high winds and sea spray, and can easily be lowered if a cyclone is threatening. These turbines can contribute a relatively high output to the commercial power network (Bal, 2001). This new technology, developed for Guadeloupe wind turbines, is now being introduced by Vergnet Océan Indien in Rodrigues and Réunion, and Mauritius has shown an interest as well. In Rodrigues, the first three turbines will generate 5% of electricity needs, and the entrepreneur's goal is to eventually supply 50% of all electricity. In Réunion, the first wind farm of 23 turbines entered into operation in 2004 in the Commune of Sainte-Rose on the southeast coast (for a total installed capacity of 6,300 kW; Vantal, 2004). A study undertaken for Région Réunion concluded that a power capacity of 135 MW could easily be installed in three specific areas of the island, for a production of 360 GWh per year, representing about 17% of Réunion's total electricity consumption (JIR, 2000).

Another local resource that should be developed in the region in the near future is geothermal energy. The technology is already available and this kind of energy has been successfully exploited in other small volcanic islands such as in Hawaii, Samoa and Guadeloupe. In Réunion, two deep exploration boreholes drilled in 1985 showed that a potential exists for discovering exploitable high-temperature resources. New exploration is underway and the principle of exploitation of the geothermal potential has been widely accepted by energy sector stakeholders. The Comoros *Initial National Communication on Climate Change* (GDE, 2002) mentions that an exploitable geothermal resource is possible in Grande Comore, but this will have to be confirmed. One geothermal deposit could provide enough energy to supply half of the island's power demand by 2020. The report also indicates possible hydro-electrical developments in Anjouan and Mohéli as well as good potential for wind power generation in Mohéli.

Concerning solar energy, all of the studied small islands show great potential due to their tropical position. Photovoltaic systems can easily service individual dwellings, farm installations, pumping stations, street lighting systems, and the lighting of commercial and office buildings. Solar energy can also be used in its thermal form for the production of hot water in solar water heaters, as well as for water desalinization. However, the price of the solar systems is still quite expensive and the cost savings are not yet evident. In Réunion, the success of their development in individual dwellings is largely due to monetary incentives provided by the public authorities. The solar option will become more and more attractive as the price of commercial power increases and the cost of solar systems reduces. Among other things, the generalisation of solar devices will reduce household demand for commercial power, either delaying the need for new power generation capacity or allowing a transfer of the saved power to other sectors, especially industry and eventually transport.

Other renewable energy sources could be exploited in the future such as biogas, domestic waste, waves and ocean thermal energy. In Réunion, the use of hydrogen is being considered as the energy necessary to produce hydrogen could be generated by wind and solar power (ARER, 2003). 'Considering the huge quantity of waste produced in Réunion and Mauritius, biogas production and waste incineration represent very interesting options as they will help resolve two problems at once (energy and waste).

Because renewable energy now offers a real option to the small islands, energy self-sufficiency is no longer science fiction. With its Regional Plan for renewable energy and rational energy use (PRERURE: Plan regional des energies renouvelables et d'utilisation rationnelle de l'énergie; RÉGION RÉUNION et al., 2003; ICE, DEBAT and INSET, 2003), Réunion is definitively taking the lead in the South-West Indian Ocean.[10] The main objective of this plan is to achieve near 100% electricity generation from renewable energy by 2025, to contribute in an exemplary manner to the limitation of greenhouse gas emissions, and while doing so, to contribute to the island's economic development as a result of the introduction of new activities. The goal of self-sufficiency in electricity will most probably not be met in 2025, but an audacious strategy and a global management of the energy sector will definitively allow the island to progress in that direction. For now, the PRERURE scenario proposes the development of an additional 372 MW in installed capacity from local energy resources (actually 225 MW: 110 MW in hydraulic and 115 MW from bagasse) that would be provided by wind power (100 MW), photovoltaic (100 MW), hydro (60 MW + 20 MW for micro hydro), bagasse (10 MW), biogas (8 MW), fuel wood (21 MW), molasses residues (13 MW), waste incineration (20 MW) and geothermal resources (20 MW).

Towards Complete Energy Self-Sufficiency?

Considering the actual forms of final energy required in the islands, and especially the importance of the transport sector needs in fossil fuels, the goal of complete energy self-sufficiency or 100% renewable energy sources supply will not be reached in the small islands of the South-West Indian Ocean for several decades. Even if we do not take into account commercial aviation and maritime activities, it will take some time before all domestic energy needs, including local transportation, are all met by local supply, if indeed this happens at all.

First, this will require a complete conversion of local transportation infrastructure to an energy fuel in which the island would be self-sufficient (electricity, hydrogen, others?), as well as the development of public transportation. Second, this will necessitate that all commercial electricity produced comes from renewable energy sources, as well as the large-scale development of small systems allowing households and other consumers (services, industry, agriculture) to generate by themselves part or all of their energy needs. Third, this will imply that a new form of energy, most probably electricity, will also replace all of the fossil fuels still used in the industrial sector. Fourth, it will require measures to develop energy efficiency and realize energy savings, and if necessary, impose limits on the energy consumption. Fifth, it will require that all of the inhibitions to progress be realised simultaneously and over quite a short period of time (at least several decades). Sixth, it will necessitate a very strong political commitment, the means (financial, technical, human) as well as a global management and a global strategy for the whole energy sector.

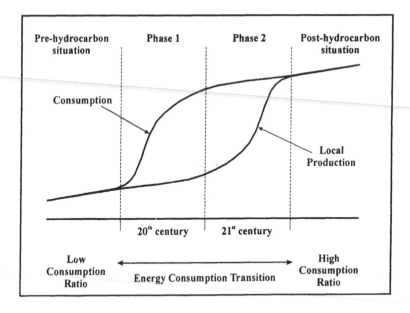

Figure 11.2. The scenario of energy consumption transition.

Nevertheless, if we consider the energy situation of the studied small islands in an historic perspective, this brings us to propose the concept of the energy consumption transition (Figure 11.2). We can effectively see that, at the end of the nineteenth century, these islands had an energy regime that can be referred to as the pre-hydrocarbon regime, characterised by a low ratio of energy consumption by inhabitant and energy self-sufficiency. Among others, the development of steam and eventually bunker oil in navigation, conversion from wood to coal for the trains in Réunion, the introduction of automobiles and trucks, the development of aviation, the development of hydrocarbon needs in industry, as well as the commissioning of thermal power stations to generate electricity, have all drastically changed this initial situation throughout the twentieth century. The result of this evolution, accompanied by the modernisation of the household equipment, is a very strong increase in the ratio of energy consumption and the development of considerable energy dependence (phase 1 of the transition).

Now, for the current century, it is possible to foresee a progressive reduction in the ratio of energy consumption per inhabitant, even its stabilisation in the context of optimised energy control, coupled with a growing exploitation of the renewable energies which will lead to the progressive replacement of fossil combustibles by locally-produced energy in transportation (at least local transportation) as well as in their other uses. At the end of this evolution that is slowly but certainly beginning (phase 2 of the transition), the studied small islands will possess a new energy regime that can be referred to as the post-hydrocarbon regime. This is characterized by a high energy consumption ratio per inhabitant and a return to energy self-sufficiency. In order for this new energy regime to eventuate, it will likely be driven by external factors (technology innovations, replacements and transfers and evolutions in the international energy market). However, the islands will definitively have a say in the progress of the energy consumption transition. Indeed, the transition could be greatly favoured by sound choices, efficient policies and management as well as voluntary actions in the energy sector. One can reasonably predict that all local transportation will not run on fossil combustibles any time before the end of this century, and, taking into account that the possibility to generate all electricity from renewable energy is absolutely real, the scenario of the energy consumption transition can therefore definitively be completed by the end of the twenty-first century. At that time, it is possible that the only remaining fossil combustibles used in the studied small islands will be in aviation and maritime transport (if a major technological breakthrough has not already changed the situation in these last two sectors).

Conclusion

If, in the past, the fossil fuel solution was technically easy and economically rational, it now appears that the wind has changed and that, in the context of the small islands, it is now globally more costly than the renewable energy solution. This is not more expensive and its environmental impact is limited, especially when compared to fossil fuel combustion. Its large-scale development will also contribute to a reduction in energy dependency and thus the energy and economic vulnerability of the islands. Coupled with increased energy efficiency and the conversion of the fossil fuel systems (industries, transportation, power plants, etc.), the renewable energy solution has the potential to deliver durable energy self-sufficiency to the islands, thus making them totally energy secure.

The mutation from their actual situation of "energy poor territories heavily dependent upon fossil fuel imports" to that of "energy rich and completely free of fossil fuel utilisation" will certainly take time. However, it is possible to progress quite rapidly if there is a strong local political commitment, with the participation of all the energy sector stakeholders in a global management framework as well as with the development and the implementation of a very efficient global strategy.

Our survey shows that only a small part of local renewable energy resources has yet been exploited in the studied small islands. This is largely due to their relatively high cost, the lack of technically-adapted systems, the lack of local expertise and the driving influence of the companies having success in and with the fossil fuel solution. Now the situation is definitively changing, as renewable energy solutions are affordable and better adapted, the know-how in acquisition is better, and the public administration is willing to take a leading role in energy management. New technological breakthroughs, especially in transportation and energy production systems, as well as national (France for Réunion and Mayotte), European and international financial support will facilitate the movement towards the large-scale exploitation of local energy resources.

Finally, regional and international cooperation will also play an important role is this venture. Due to their remoteness, their small population and limited means, the sharing of different experiences and a collective approach is particularly desirable. At the international level, energy has been selected as one of the fourteen priority fields for small island developing states (UN, 1994, Barbados Plan of Action) and a Global Sustainable Energy Islands Initiative (GSEII) has been organized. At the European level, the EU has developed a strategy "Towards 100% Renewable Energy Supply" for its island territories, an initiative that is complemented by the Island 2010 campaign (INSULA, 2001). At the South-West Indian Ocean level, collective action in the field of energy is now developing, while the first regional seminar on "Renewable energies for the sustainable development of the French-speaking Indian Ocean countries" was organised in March 2004 in Mauritius[11] (IEPF et al., 2004). If there is not yet an official cooperation programme on energy issues by the Commission of the Indian Ocean (COI), the recently-founded University of the Indian Ocean is already developing the expertise in that field.[12] To date, regional cooperation in the energy sector has been primarily undertaken by the private sector. Now scientists, engineers and public officers from all of the islands (including Madagascar) are joining in regional networking on energy issues.

Therefore, the conditions are in place for a real energy revolution in the small islands of the South-West of the Indian Ocean. However, if significant and efficient actions are not taken very rapidly to develop renewable energy exploitation and maximisation of the use of energy, this will lead to a worsening of the regional energy balance and to an increase in regional energy dependency. Nevertheless, it is very likely that the situation will evolve in a positive manner and not to a distant future (before the middle of the century) with the generalisation of local energy exploitation and the eventual replacement of fossil fuels in local transportation.

Acknowledgement

Special thanks to M. Léo L. Larivière, Technologist for the Department of Geography at Laurentian University, for the drawing of the two figures.

References

ARER, 2003: *Énergie Réunion - Enjeu Réunion*. Agence régionale de l'énergie Réunion, Saint-Pierre (Réunion). http://www.arer.org/download/endurable/politique/panorama_enjeux_energie_reunion_arer.pdf (last visited on December 20th, 2004).

Bal, J.L. et al., 2001: The development of renewable energy sources for electricity generation: the example of the French Overseas Departments and Corsica. *Towards 100% RES Supply: Renewable Energy Sources for Island Sustainable Development*. International Scientific Council for Island Development (INSULA), Paris.

CIA, 2004: *The World Factbook 2004*. Central Information Agency, Washington. On the Web at: http://www.cia.gov/cia/publications/factbook/index.html (last visited on December 20th, 2004).

CPI-Réunion, Internet: «Section sur les filières énergétiques : Situation énergétique à la Réunion». Comité de pilotage de l'industrie de l'île de la Réunion: http://www.cpi.asso.fr/version2/energies01.htm (last visited on October 28th, 2004)

CSO, 2003: *Digest of Energy & Water Statistics - 2002*. Central Statistics Office, Port-Louis.

CSO, 2004: *Energy & Water Statistics - 2003*. Port Louis : Central Statistics Office, Ministry of Economic Development, Financial Services and Corporate Affairs, 19 p.

EIA, Internet: «International Homepage: Country reports». Energy Information Administration, Washington (D.C.). http://eia.doe.gov/ (last visited on October 28th, 2004).

GDE, 2003: *Initial National Communication on Climate Change* [for the Comoros]. Union of the Comoros, Ministry of Development, Infrastructures, Posts and Telecommunication and International Transports, General Directorate of Environment, Moroni.

Godard, Henry (Coordination), 1998: *Les Outre-mers*. Montpellier / Paris: GIP Reclus / La Documentation Française. Atlas de France, volume 13. 128 pages.

ICE, DEBAT and INSET, 2003: *Mission d'assistance à la réalisation d'un plan énergétique régional, volume complémentaire 1 : prospective des consommations d'énergie*. Saint-Denis : Conseil Régional de La Réunion, Marché no. 20020119, 77 pages.

IEPF et al., 2004. *Les énergies renouvelables pour le développement durable des pays francophones de l'Océan Indien"* [Actes du Séminaire régional, Port-Louis, Île Maurice, 1er- 4 mars 2004]. http://www.iepf.org/docs/publication/maurice.pdf (last visited on December 28th, 2004).

INSULA, 2001: *Towards 100% RES Supply: Renewable Energy Sources for Island Sustainable Development*. INSULA: International Scientific Council for Island Development. On the Web at: http://www.islandsonline.org/island2010/ (last visited on May 20th, 2004).

INSEE Réunion, 1999: *Tableau économique de la Réunion 2000*. Institut National de la Statistique et des Études Économiques (INSEE), Direction Régionale de La Réunion, Saint-Denis (La Réunion).

INSEE Réunion, 2003: *Tableau économique de la Réunion 2003/2004*. Institut National de la Statistique et des Études Économiques (INSEE), Direction Régionale de La Réunion, Saint-Denis (La Réunion).

INSEE Mayotte, 2003: *Tableau économique de Mayotte 2003/2004*. Institut National de la Statistique et des Études Économiques (INSEE), Antenne de Mayotte. Mamoudzou (Mayotte).

JIR, 2000: «La Région complète l'étude sur le potentiel éolien de La Réunion». *Journal de l'île de la Réunion* (newspaper article published on August 24, 2000), Saint-Denis (La Réunion).

Juliette, Yvon Calix, 2004. «Utilisation des énergies renouvelables aux Seychelles» in IEPF et al. (2004), *Les énergies renouvelables pour le développement durable des pays francophones de l'Océan Indien*, pp. 23-25.

MISD, 2003: *Seychelles in Figures, 2003 Edition*. Ministry of Information Technology and Communication (MITC), Management and Information Systems Division, Victoria.

PRB, 2004: *2004 World Population Data Sheet*. Population Reference Bureau, Washington, D.C. http://www.prb.org/pdf04/04WorldDataSheet_Eng.pdf (Last visited on December 20th, 2004).

RÉGION RÉUNION et al., 2003: *PRERURE : notre avenir énergétique est entre nos mains* [plaquette d'information]. Région Réunion et partenaires (ADEME, EDF, ARER et UE), Saint-Denis (Réunion).

UN, 1994: *Report of the Global Conference on the Sustainable Development of Small Island Developing States* (Bridgetown, Barbados, 25 April-6 May 1994). United Nations, General Assembly, A/CONF.167/9, New York.

UNDP, 2004: *Human Development Report 2004*. The United Nations Development Programme (UNDP) and Oxford University Press, New York and Oxford. http://hdr.undp.org/reports/global/2004/ (Last visited on December 20th, 2004).

Vantal, Anne, 2004: «Énergies renouvelables : rendez-vous à la Réunion». *Systèmes solaires*, no. 159, dossier of 16 pages.

VIRTUAL SEYCHELLES, 2003: Webpage entitled "Infrastructure: Electricity Supply". http://www.virtualseychelles.sc/busi/busi_infra_electric.htm (Last visited on December 20th, 2004).

NOTES

1. "There are many disadvantages that derive from small size, which
    are magnified by the fact that many island States are not only small
    but are themselves made up of a number of small islands. Those
    disadvantages include a narrow range of resources, which forces
    undue specialization; excessive dependence on international trade
    and hence vulnerability to global developments; high population
    density, which increases the pressure on already limited resources;
    overuse of resources and premature depletion; relatively small
    watersheds and threatened supplies of fresh water; costly public
    administration and infrastructure, including transportation and
    communication; and limited institutional capacities and domestic
    markets, which are too small to provide significant scale economies,
    while their limited export volumes, sometimes from remote
    locations, lead to high freight costs and reduced competitiveness.
    Small islands tend to have high degrees of endemism and levels of
    biodiversity, but the relatively small numbers of the various species
    impose high risks of extinction and create a need for protection"
    (UN, 1994, *Programme of Action for the Sustainable Development of
    Small Island Developing States*, Preamble, par. 4).

2. Some commercially exploitable offshore oil deposits may be found
    on the Seychelles Plateau. Several preliminary exploration
    campaigns have already taken place and seem to confirm this
    potential. However, no new exploration is underway and
    exploitation seems to be very far away.

3. Developed by UNDP in the beginning of the 1990s, the concept of
    human development postulates that the more a population has a
    long and healthy life, the more it is educated and the more it
    disposes of financial resources, the more its human development is
    high. The human development index (HDI) is a summary measure
    of human development.

4. One tonne of oil equivalent (toe) is defined as having a calorific
    value of 397 Therms (1 Therm = 25,200 kilocalories). According to
    INSEE Réunion (2003, p. 174), 1,000 kWh represents 0.086 toe, 1
    tonne of coal represents 0.669, 1 tonne of jet fuel or of gasoline
    represents 1.048 toe, 1 tonne of liquefied petroleum gas (LPG)
    represents 1.095 toe. There is about 7.3 barrels of crude oil per metric
    tonne.

5. As aviation and maritime activities (transport and fisheries) are
    intrinsically related to the island system, thus their energy
    consumption has to be included in the island total energy needs.

6. France official factor of conversion for electricity to toe has changed
    in 2002 and Réunion Island figures for 2000 use the new value.
    Before, 1,000 kWh of hydro or solar generated electricity was

considered to represent 0.222 toe. Now, the same 1,000 kWh is considered to represent only 0.086 toe.

7. Seychelles Petroleum (Sepec) is also servicing the many large international fishing fleets based in and beyond the Seychelles and began tanker activities to serve neighbouring coastal African communities.

8. "The sugar cane industry produces a residue called bagasse, which is the fiber of the cane after sugar has been extracted. One metric tonne of cane produces about 320 kg of bagasse. Bagasse has a net calorific value (NCV) of 7,900 kJ/kg which is greater than the NCV of many lignites mined in the world very expensively." (Bal et al., 2001)

9. At the beginning of 2004, some 47,000 solar water heaters were in operation in Réunion (Vantal, 2004, p.7).

10. As a French Overseas Department, an European Island and an Ultraperipheral Region of the European Union, Réunion beneficiates of an important financial support for its development. France government and European Union will finance part of the projects related to the PRERURE.

11. Organised under the leadership of the *Institut de l'énergie et de l'environnement de la Francophonie* (IEPF, based in Quebec City, Canada), in partnership with *Fondation Énergies pour le Monde* (FONDEM, France), *Région wallonne de Belgique* (Belgium), *Agence de l'environnement et de la maîtrise de l'énergie* (ADEME, France) and Mauritius Ministry of Environment and National Development Unit, this seminar was attended by policy-makers and national energy specialists from the Comoros, Madagascar, Mauritius, Réunion and Seychelles.

12. The University of the Indian Ocean (U.O.I.) organizes graduate-level formation on "Economy and energy planning with environment preservation" (EPEE).

ENVIRONMENTAL SECURITY AND ENERGY SECURITY

CHAPTER 12

The Indian Ocean as *The Nuclear Ocean:* Environmental Security Dimensions of Nuclear Power

Timothy Doyle

Introduction: The Securitisation of the Environment

The emergence of environmental security is sometimes referred to as the 'securitisation of the environment'. For those interested in promoting environmental solutions, there are positives and negatives involved in this reframing of *the environment* in this manner. On the one hand, environment is elevated to the upper echelons of p⁰ licy-making in, and between nation-states. Security, identity and economics continually occupy these high-level portfolios, and their lofty positions in governmental decision-making is usually matched by high levels of funding and power. Some environmentalists argue that their inclusion in these powerful security roundtables and other policy-making mechanisms provides similar access to such finances and power. On the other hand, there is a danger that the rhetoric of environmental change is simultaneously co-opted by more mainstream, traditional security interests, with this co-option leading to eventual disempowerment and displacement of environmental goals by others.

The definition of environmental security, as espoused by the IORG, is an inclusive one: a positive definition of green security and sustainability; one that transcends nation-state boundaries; and relates to conditions which secure individual access to a basic infrastructure for survival in a geopolitical region defined by shared environmental boundaries. Environmental security, in this vein, is reliant upon shared understandings of ecological conditions leading to potential and real conflicts, as well as developing a more sustained, peaceful, and resource-secure regional future (See Doyle 2004a).

Despite the varied rhetoric, let me briefly explain what is no doubt the key attribute of all environmental security issues, for not all environmental issues constitute environmental security concerns. First and foremost, it is critical to understand that environmental security goes beyond the national interest; that it relates to a broader level of human security: in the case of the IORG, this can be understood by envisioning shared problems and issues at the regional level. Usually, these are not issues that exclusively relate to sub-national conflicts, although these form an important part of the larger whole. Sub-national environmental issues primarily remain the core business of nation-states.

Climate change and population growth are two of the most readily recognisable environmental security issues which have emerged since the Earth Summit, in Rio de Janeiro in 1992. These readily cross the boundaries of nation-states. As far as the Indian Ocean Region is concerned, they are important issues; but there is no doubting that they have emerged on the green security agenda largely as a result of northern initiatives, rather than reflecting critical environmental needs of the majority of the people living in the IOR. There are obvious historical reasons for the domination of the North. At the end of the 1980s, when these agendas for common futures were first being drawn up, predominantly *northern* issues were being recast as *global* ones. With the turn of the millennium, however, *southern* environmentalists are now shaping their own green agendas far more, and this has had a concomitant impact on the shape and substance of environmental security agendas, with less emphasis on, for example, climate change, and more priority given to basic survival issues such as shelter, nutrition and opposition to wars.

A concept of environmental security which is more inclusive of the interests of the majority of people in the Indian Ocean states, both littoral and non-littoral, is one that moves away 'from viewing environmental stress as an additional threat within the (traditional) conflictual, statist framework, to placing environmental change at the centre of cooperative models of global security' (Dabelko and Dabelko 1995, 4). But to do this, there must be an increased understanding of the environment, not as an external enemy force but as a diverse nature which is inclusive of people; a nature which has the potential to provide secure access to individual citizens of all countries in the Indian Ocean Region to basic nutrition, adequate access to healthy environments; appropriate shelter, and, the security to practice a diverse range of livelihoods which are both culturally and ecologically determined (Doyle 2004b).

Nuclear Issues as Environmental Security

Like climate change and population growth, nuclear issues cross nation-state boundaries with ease. The globe is at an interesting point in its relation to nuclear power. Despite the recent re-championing of nuclear power by the Bush administration in the United States, many industrialised countries are reducing their reliance on nuclear energy, recognising its inherent, long-term environmental dangers. Wealthy countries, such as the heavily industrialised Germany, are now in the process of decommissioning their nuclear industry over the next thirty years. This has brought its own problems, and the case of Germany – reviewed later in this Chapter – provides a future snapshot for other states as it wrestles with the environmental security issues which it is now saddled with, as it seeks an end solution to the transport and storage of its vast deposits of dangerous nuclear waste.

Just as many parts of the industrialised world have come to the realisation that the nuclear option cannot be sustained in an environmentally secure fashion, parts of the majority world have now embraced the technology. The Indian Ocean Region (IOR) is now undergoing rapid nuclearisation, with many countries inhabiting the IOR now embracing the technology for the first time. It was most appropriate that this chapter was first presented as a paper in Tehran, as Iran, also, now seeks to become a "nuclear state"; but, like so many of its Indian Ocean neighbours, the issues of nuclear waste, its safe transport and its final disposal have not been high on its environmental or nuclear agendas.

Issues pertaining to nuclear power have so far largely escaped being included in debates about environmental security. One of the key reasons behind this is that nuclear issues are already considered as part of the mainstream, traditional, hard security or defence debates. This focus on nuclear energy is almost always from the perspective of weaponry and arms proliferation, and is almost always centred on the identities of nation-states. So, it is not so much introducing nuclear issues into the security rhetoric out of the policy cold; but rather, dragging some of its related issues out of "traditional" security debates, and then including them in the more alternative, human and environmental security discourses.

For the purpose of this Chapter, I have not purposively entered into the debate over nuclear weapons build-up in the region (which, of course, is already high on the international security policy agenda); but, instead, I have sought to recast the nuclear question in the Indian Ocean Region as an environmental security issue. Two key questions can emerge by focusing on two areas of the nuclear fuel cycle: i) dealing with unintended releases of radioactive materials from reactors; and ii) dealing with the ever-increasing problem of nuclear waste, its transport and its storage.

The first question is more obvious. In what is one of the few attempts to address nuclear energy issues as an environmental security issue, an excellent paper by Barnett and Dovers, (though with a Pacific focus) provides a nice framework through which to view such issues. The authors list six parameters of environmental security: i) the spatial scale of causes and effects; ii) the magnitude of possible impacts; iii) the temporal scale of possible impacts; iv) the reversibility of impacts; v) the measurability of factors and processes; and vi) the degree of complexity and connectivity (Barnett and Dovers, 2001). Environmental security issues, whether they pertain to the dangers of reactor meltdown, accidental omissions, or release of spent/re-processed fuels during transport/storage, measure highly on the scale in all six categories.

First, the spatial scale of an accidental release of nuclear material, either from a reactor meltdown, or leakage; or from an accident associated with the transport and disposal of waste, are immense. At this level, they are instantly a problem at the regional level. For example, between 1952 and 1986, there were 14 reactor accidents involving core damage; but the Chernobyl accident was by far the most severe. Environmental security issues spread across a vast spatial scale, crossing nation-state boundaries. At Chernobyl, long-lived radioisotopes Caesium–137 and Strontium–90 spread over 100,000 kilometres (Marples 1993, 39). In the warmer waters of much of the Indian Ocean this spatial spread would usually been even greater, due to the higher levels of rainfall. The magnitude of impacts also rates highly. Barnett and Dovers write:

> Some 20-25 per cent of the annual budgets of the Ukraine and Belarus are now spent on coping with the consequences of the Chernobyl accident, although some consider the economic costs to be unbounded. In health terms, some of the populations exposed to fallout from the Chernobyl accident have experienced a hundred-fold increase in childhood thyroid cancers. Tellingly in terms of security, the magnitude of the clean up after Chernobyl has been likened to the task of rebuilding after the German-Soviet war (Barnett and Dovers, 2001, 164).

In this light, the safe production of energy is not a luxury issue pertaining to the first world, but a necessary condition for survival in all parts of the globe. The third characteristic – the temporal scale – is obvious. Fifty to sixty per cent of the reactor's inventory of Iodine 131 was released causing massive increases in thyroid cancer. Also, other longer lasting radioisotopes were released – strontium-90 and caesium-137 – both will remain intensely harmful to human health for at least 300 years (Clayton et al, 1986). Whilst there is much talk of the welfare of future generations in the rhetoric of sustainable development, it must be understood that even medium level nuclear waste has a half shelf life of thousands of years.

Nuclear accidents also qualify under the remaining three categories and their impacts are not reversible, with the permeation of isotopes into the soil, water, plants and animals leading to pervasive contamination. Also, they are highly immensurable and are impossibly complex (Barnett and Dovers 2001, 164). The environmental security concerns of unintended releases of nuclear material, as Barnett and Dovers rightly argue, are obvious, if not neglected, in the Indian Ocean context.

Before this Chapter can concentrate on its core task – oration of the ever-increasing problems of the storage of nuclear waste and its transport in the IOR – it must be fully understood that opposition to nuclear power in the region - to this point in time – has been minimal, except in the case of Australia. The reasons for this are many. First of all, the majority of the Indian Ocean is third world, and nuclearisation has normally been a characteristic of industrialisation. However, most importantly, the dangers of nuclear accidents – whether they occur during plutonium production or during the handling and storage of waste products – are often regarded by majority world governments as constituting acceptable levels of risk. In the current case of Iran, with fears of a US-led invasion dominant in its policy circles, the possibilities of a nuclear accident are not rated high on either security or environmental agendas. Concerns over such environment risks are often regarded as *luxury* concerns.

At the other end of the Indian Ocean, Australia, on the contrary, has been at the vanguard of anti-nuclear movements in the global theatre, along with the aforementioned Germany. Some of the questions on environmental security which have emerged from the Australian and German experiences have informed the more affluent, minority world, and these arguments are a harbinger of what is to come in the Indian Ocean region as a whole, as it becomes *the nuclear ocean*.

A Brief History of Nuclear Issues and Movements in the Minority World

Anti-nuclear movements emerged across the globe in the years subsequent to the bombing of Japan in World War II. Peace movements have, for a long time, often been one and the same with anti-nuclear movements. The nuclear clouds over Hiroshima and Nagasaki have invoked images of a nuclear holocaust for all who have lived since. During the Cold War between the ex-USSR and the United States, it seemed to many that a modern, industrial Armageddon was imminent. In the days of the Reagan administration in the US, and of Gorbochev in the USSR, the doomsday clock clicked 'one minute to midnight'. During this period, in surveys of teenagers in the West, one of the most commonly shared dreams of the survey participants was of global nuclear destruction (Doyle, 2005).

Apart from fears of global devastation from nuclear weaponry, anti-nuclear movements have also rallied against the production of nuclear energy. They have argued vociferously that the processes of transition from uranium ore to plutonium; its storage; and then ultimate disposal of nuclear waste products are inherently dangerous. There have been countless instances where nuclear *meltdowns* and other accidents have occurred at reactors which had been reported to be 'failsafe'. An excellent example occurred at the Three Mile Island nuclear power plant in Harrisburg, in the United States, in 1979. Radioactive gases were released, and this was followed by an evacuation of the surrounding area. An equally notorious incident, as already mentioned, occurred at the Chernobyl nuclear power plant near Kiev in the Ukraine. Over 130, 000 inhabitants had to be evacuated from the region. Only the future will tell us of the final death count, although 30 died directly from the meltdown, and hundreds of others were treated for severe radiation sickness (Papadakis, 1998, 41). Many Europeans, downwind from the reactor explosion, were lied to by government officials, who justified their silence on the basis that they did not wish to 'panic' the population. In Austria, for example, days after the event, citizens were finally told to stay indoors and close their windows. For months afterwards, fresh fruit and vegetables were off many European's menus.

There can be no doubt that there have always been enormous questions as to the economic viability of nuclear energy. By assessing it on purely on an economic basis, many European states, at the beginning of the third millennium, are now questioning, and even attempting to phase out, their nuclear programmes. The reasons behind advocating nuclear power, however, have not always been about the economically sustainable production of energy. Much support for the pro-nuclear lobby comes from industries devoted to the building and supply of weapons used in advanced, global warfare. These industries are often inseparable from the 'defence' interests of nation-states. No clearer an example need be sought than to refer to the bombing of Greenpeace's flagship, *The Rainbow Warrior*, as it stood moored in Auckland Harbour in 1984. After intensive investigations, it was revealed that this destruction of human life and property was instigated by French secret service agents working on behalf of the French Government to end protests against their nuclear weapons detonations in the Pacific.

After a gradual waning of the nuclear industry in the first world, the election of George W. Bush as President of the United States in 2001 has reinvigorated support for the nuclear industry through such programmes as the United States 'missile defence strategy'. Also, large military industrial corporations are now targeting third world states, such as India, Pakistan, and Korea and Iran, in a bid to sell their wares. As mentioned at this chapter's outset, with the birth of a nuclear industry in these countries, there is also a concomitant emergence of majority world anti-nuclear movements, for the first time (Bidwai and Vanaik, 2000).

The anti-nuclear movement is very much part of the *political ecological* tradition of environmental movements. Apart from fighting against the specifics of the nuclear industry from mining, to plutonium production, to nuclear waste disposal, the movement has always been at the radical vanguard of green movements across the globe. Its very raison d'être challenges the concept of progress at all costs; it attacks the concept of short-term gains and short-term greed; it assails the concept that science and *end-of-pipe* technologies can make environmental risks simply disappear; and it objects to the overly close connections between the state and big business, often referred by movement participants as the *military-industrial complex*. Although it has often dabbled in the worlds of political parties and movement organisations, nuclear movements have continued to provide *counter-cultural* dimensions to many national environmental movements. Again, this more radical tradition of the anti-nuclear movement has been suited to the first world, as its primary focus is an attack on advanced industrialisation. As the majority world – and in this instance, the IOR - becomes increasingly industrialised, such arguments will gain more currency.

As aforesaid, two of the most forceful and sustained anti-nuclear movements have emerged in Germany and an IOR country, Australia. Nuclear protest began early in Germany. In 1975, protests at Whyl began as small communities were concerned about the impact of nuclear power production on their vineyards. These protests quickly became national, and "escalated into a widespread antagonism toward three things: the collusion between the state government of Baden-Wurttemberg and industrial interests, the deployment of the police to solve political problems, and the dominant political parties" (Papadakis 1998: 39). Throughout the mid to late 1970s protests continued, culminating in the single biggest political rally ever held in West Germany, at Bonn, in October 1979, where 150,000 people gathered in protest over the nuclear industry. Right from the beginning, the anti-nuclear movement was at the centre of the formation of Germany's modern environmental movement. Papadakis supports this contention when he writes:

> During the same month, over 1,000 environmentalists met in Offenbach to discuss the possibility of establishing a new political organisation that later became known as Die Grünen (Papadakis, 1998, 40).

At a very similar time, the Australian anti-nuclear movement was developing. Although there was no dramatic building programme of nuclear reactors to rail against, as was the case in Germany, Australia was in the process of becoming one of the world's largest uranium producers and exporters. In 1975, Friends of the Earth (FoE) became increasingly vocal and active in their anti-nuclear campaign (Doyle, 2000, 133). In 1976 and 1977 up to 50,000 people gathered in mass rallies in Australian cities, and, in June 1997, anti-uranium protestors tried to stop the ship *ACT 6* from exporting yellowcake from the Lucas Height reactor: 40 arrests ensued. Until this time of writing Australia still only has one nuclear reactor, at Lucas Heights, though there is currently much pressure to upgrade this facility. In a short history of the Australian Peace Movement, from which the anti-nuclear movement, in part, evolved, Summy and Saunders explain the reasoning behind these early mass protests:

> Their objections were not confined to the link between nuclear weapons proliferation and the expansion of the

nuclear power industry, but were based on a host of other factors as well. These included the environmental hazards of nuclear-waste disposal, reactor accidents, and releases of radioactivity at various stages of the nuclear fuel cycle. Other problems raised included the social and political threats to a society from terrorist use of nuclear materials, from reduction of civil liberties, and from the centralisation of political and economic controls in the hands of financial bureaucratic elites (Summy and Saunders, 1986, 45).

Current Nuclear Issues in the Indian Ocean: The Case of Australia

Nuclear reactors and nuclear waste dumps are either being constructed, or are under consideration, for the first time in many parts of the IOR. Australia is currently in the early stages of building a new nuclear reactor to replace its old research reactor at Lucas Heights in Sydney. Other IOR states, including India, Pakistan, Iran and South Africa, have all either recently developed a nuclear programme, or are in the initial stages of development. Linked to these reactors is the need to consider more permanent repositories for nuclear waste, as well as the issues which arise out of the transport of such waste. Let us take a snapshot of the anti-nuclear movement in Australia. At this time of writing, there are at least five major anti-nuclear campaigns unfolding concurrently:

1. Jabiluka

The most world-renowned Australian anti-nuclear campaign is the Jabiluka campaign fought at Kakadu in the Northern Territory. Its arguments are three-pronged, inter-changing between the rights of indigenous peoples (which is the most dominant voice), the values of Kakadu wilderness in its national park, and its anti-nuclear stance against uranium mining. The actors in Jabiluka are largely generation Xers and local aboriginal communities. The Gen Xers were recruited by mainstream green organisations such as the Australian Conservation Foundation, mainly from large capital cities on the eastern seaboard.

2. Roxby Downs

The campaign against Western Mining Corporation's (WMC's) Roxby Downs is aimed at curtailing the expansion of one of the biggest uranium mines in the world. Mostly this campaign is run out of South Australia. Environmentalists opposing the mine primarily utilise anti-mining and anti-nuclear arguments, although there are some subservient indigenous arguments, and a very small number oriented around wilderness values. The aforementioned campaigns focusing on the Roxby Downs in South Australia are largely confined to that state. Most eco-action takes place in the capital city, Adelaide, with some direct actions occurring at the mine sites themselves and within local Aboriginal communities. These activists are usually older, and have been fighting these campaigns, specifically the expansion of Roxby, for fifteen years. They have a sound knowledge of the complete nuclear cycle, beginning with mining and finishing with nuclear waste. They have little support from mainstream non-governmental organizations (NGOs) based along the eastern seaboard of Australia, in cities such as Sydney, Melbourne and Canberra.

3. In-situ Leach Mines

Also in South Australia are campaigns levelled at halting the development of two 'in-situ leach' mines which are currently under production. The mines are owned by two transnational corporations – General Atomics (primarily a US corporation) and Southern Cross (a Canadian-based company). Most of the movement's arguments in these cases are levelled at inappropriate 'third world' technology being utilised by these companies, which pump a uranium-depleted, acid-based solution back into the aquifer when the mining process has been completed. This process is illegal in the corporations' 'home' countries. This campaign is being fought by a close association of green activists from the South Australian office of a national NGO, the Australian Conservation Foundation, alongside Aboriginal communities living near these mines. This campaign has been waged by local residents living next to the reactor in Sydney. Due to the fact that most Australians live in Sydney and Melbourne, this campaign gets support from large international NGOs like Greenpeace and Friends of the Earth.

4. Lucas Heights Reactor

In New South Wales, another Australian state, there is a key anti-nuclear campaign which focuses on shutting down the existing Lucas Heights Reactor in Sydney, and preventing the building of a new reactor facility to replace it. Also in Sydney is a campaign which focuses on matters of nuclear war. Most particularly, strategic arguments have emerged which contest the United State's proposed missile defence strategy championed by the US President George W. Bush.

5. International and National Waste Dumps

In Western Australia and South Australia large campaigns based on mass mobilisation techniques are aimed at stopping Australia, and those specific states within Australia, from becoming an international and/or national nuclear waste dump. On the other hand, campaigns against nuclear waste in Australia (reminiscent of movements against the transport of nuclear waste in Germany, see later in this chapter) involve the full gamut of people from school children to elderly pensioners; bureaucrats to business people to farmers. There is little campaign rhetoric emanating from these networks which relates to the danger of the nuclear fuel cycle; rather, it concentrates on concerns of dumping wastes in activists' localities. There are also professional campaigners who link waste issues to the generation of nuclear power, and who connect this issue with opposition to the building of the new reactor at Lucas Heights. David Noonan, of the national green NGO, the Australian Conservation Foundation, writes:

> The national radioactive dump is driven by the plan for a new Sydney reactor. To gain the public and political credibility to be allowed to build the proposed $500 million reactor the Australian Nuclear Science and Technology Organisation (ANSTO) needs to clear this site of radioactive wastes. This is planned to occur by transport of lower level wastes to the repository in central South Australia, and

transporting the high level spent fuel rods across the oceans to France and Argentina for reprocessing. With resultant reprocessed nuclear waste to later be returned to Australia through some as yet unnamed port and then transported onto the store (Noonan, 2000, 1).

In all of these campaigns there are enormous differences and some similarities which go far beyond the form and content of argumentation. The most blatant similarity is that most participants see themselves as part of the anti-nuclear movement, and many also see the anti-nuclear movement as a sub-set of the environmental movement, though not all. The differences are more obvious, more numerous and more profound. In most cases, different networks of activists dominate. Partly as a consequence and partly as a causational factor, each network possesses different goals and advocates different strategies for ascertaining those goals. This is caused by and results in very separate demographic profiles for activists involved for each campaign (Doyle 2005).

Most of these campaigns in the Australian context are either aimed at halting the nuclear fuel cycle at its base: that is, to challenge the validity of mining uranium ore itself; and to object to the construction of a storage facility for the waste products of these processes. In this manner, opposition against nuclear programmes is firmly placed within the discourse of environmental security as laid out earlier in this chapter.

Transport and Storage as an Environmental Security Issue

In Australia – the country with the most developed opposition to nuclear energy in the IOR – the next level of campaign will be aimed at the proposed transport of nuclear waste, once (and if) plans for a centralised waste facility becomes a reality. Already, however, anti-nuclear activists, both within mainstream parties and within the broader social movements, are building a new case for opposition based purely on environmental security issues which will inevitably emerge from the transport of such wastes. The transport of nuclear waste crosses both state, national and international boundaries: it comprises an environmental security issue in the purest sense. Within the IOR, there are two key areas of concern. The first relates to radioactive waste being transported by sea from international reprocessing units in Europe. The second area relates to the transport of waste by land. This may include waste which emerges from the mining process (reviewed using the case of India, in the next section), as well as the more toxic substances which evolve after the processes of nuclear reaction. As far as the latter scenario is concerned, this has not yet occurred in sufficient quantities in the IOR. Again, the case of Germany, now wracked with huge and extremely expensive environmental security issues relating to the safe transport and storage of waste, provides IOR nations with a taste of the future. As a consequence, before we visit the current realities of ocean transport across and within the IOR, let us briefly review the German situation in relation to land-based transport.

The Battles for Gorleben: the Case of Germany

In the early 1980s, in the Northern German farm town of Gorleben, a huge hall was designated as an 'interim' depository for high-level radioactive waste. It stood empty for twelve years until April 1995, when the first castor (a cask used for waste storage) was delivered (Die Tageszeitung, 1995, 24-29 April). Demonstrations began several days earlier when 4,000 people gathered in the streets of Gorleben. The waste had originally made its way from the reprocessing plant at La Hague in France. Germany has no such facility, nor have many other nuclear states, for treating its own waste. Put succinctly, the reprocessing system in France takes quantities of spent fuel rods from countries with reactors, and increases the mass of the waste by approximately 30 times; but simultaneously decreases the *shelf life* of the waste, so that it may be stored more *safely*. Regardless of this *reprocessing,* the waste is still regarded by international standards as *high-level* radioactive waste.

On the Monday after these initial protests, the transport began again on German soil from a nuclear power plant in Phillipsburg in southern Germany. As soon as the transport of the waste began, protesters blocked its every move. At first they attempted to block the loading of the waste onto the train: the Deutsche Bahn AG Railway. On the following morning, a massive 2,500 people tried to prevent the conveyance of the waste after it had been unloaded from the train onto road transport. During this protest, as large as the numbers were, the eco-activists were greatly outnumbered by approximately 7,600 policemen.

The cost of the police deployment, amongst other things, reportedly cost the German Government about $US18 million. The second shipment of waste which took place a year later cost more than $US50,000, with an estimated 9,000 protestors and 15,000 police. Incredulously, a third transport was attempted in March 1997. During this movement of waste at least 20,000 protestors took to the streets, with an official figure of 30,000 police also in attendance. According to estimates, the movement of eight castors of waste cost the German Government and its nuclear industry $US180 million (Mariotte, 1997, 3). It seemed after the third shipment that the activists had won the day, as they had deliberately set out to make transport of the nuclear waste economically non-viable. As a partial outcome of these protests, the then CDU federal government temporarily suspended further shipments.

The opposition to the shipments has been widespread on all three occasions. In personal interviews with anti-nuclear activists involved in the Gorleben protests, what emerged, perhaps unsurprisingly, was the immensely broad demographic spread of the activists. In one such interview, an activist describes the participants as follows:

> I would say there is a mixture of older and younger people. Quite a lot of people in their 30s, some veterans from the 70s and 80s, and quite a few younger people. I know a couple in their 70s who usually take part in the protests . . . In particular, among the people of Wendland, including the farmers who have really been at the forefront of the movement, you find people of all ages and classes (Skullerud in correspondence with the author, 2001).

This anecdotal evidence is further supported by other, more academic sources. Blowers and Lowry (1997, 51) also contend that the protests included local opposition organised by the *Burgerinitiativ*, a citizen-based organisation, as well as external, established green organisations at both the lander and national level, such as Greenpeace. The reasons behind the protests, as often the case with environmental coalitions, were multifarious. Local farmers and businesspeople were concerned about the health and safety risks to their produce and to their communities, whilst the established green organisations were 'often dancing to a different drum, more radical and intent on the destruction of the nuclear industry'.

The first and second protests were non-violent. Anti-nuclear participants stood or sat in front of the trucks. Regardless of this low-level of actual threat to police, on both occasions, police used water cannons to disperse the protestors. In cold weather, the use of waster cannons can be very severe on the recipient. During the third event, although it always aspired to be non-violent, there is some evidence as to the increasing militancy of the protests. One eye-witness account describes some of the alternative tactics employed:

> During the night, a group of Autonomen – radicals perhaps not organised enough to be anarchists – engaged in serious street fighting near the town of Quickbon, along the only road the casks can now travel. That road hasn't been torn up, and the radicals want time to set up barricades and dig under the road . . . Back to the present. In the centre of the village, about 80 tractors – some chained together – form a barricade. A kilometre and a half down the road, there are more barricades – downed trees, dirt, concrete, and whatever the locals could find. Underneath the barricades, the road has been completely dug out – just a few centimetres of road and then holes several meters deep. If the 90-tonne casks try to move down this road, it will collapse. There are only farmers and workers in this village. The students and anarchists often associated with German anti-nuclear protests are elsewhere. We go to a farm which has been set up as an impromptu information centre and resting station. Above us, police helicopters circled constantly, the noise of the blades now so common they're just part of the background (Mariotte, 1997, 5).

In late March 2001, another battle for Gorleben took place. Another six castors were taken to their destination; another enormous amount of money spent in order to protect the deadly waste. On this occasion, 20,000 strong riot police were present, turning the water cannon on thousands of eco-activists. What is most interesting about the German case, for member states of the IOR, is that opposition to nuclear energy was focused on its place of transport. Attacks on the place of transport will one day become a major environmental security issue for all new nuclear powers, including those in the IOR. Of course, it is these places and routes of transport which, apart from being vulnerable to direct action protests, may also prove to be the sites of environmental terrorism.

Transport and Storage of Nuclear Waste Within the Indian Ocean

As discussed at length, traditionally, the key problem for the globe in relation to waste has been waste disposal *after* atomic reaction occurred. However, because most countries in the IOR region have infant nuclear programmes, this problem has not yet reached its potential magnitude. However, what is already pressing in these less affluent countries, are issues of waste management which relate directly to the mining process itself. In wealthier countries such as Australia, the management of these wastes in tailings dams is usually based on adequate levels of technology, although there are still dangerous leakages reported at disturbingly regular intervals. In the poorer parts of the region, these storage issues pertaining to tailings dams are further exacerbated by the increased levels of environmental risk which are considered by both state and corporate players to be acceptable.

Let us consider the case of India. India has embarked on a three-stage nuclear energy programme which its government believes will reach its full fruition by 2020, with 20,0000 mw of electricity produced. One of the key problems for India relates to its very low grade uranium stocks, thus necessitating mining on an extremely large scale, with few margins available to pursue adequate environmental and safety measures. As a consequence, 'normal' safety standards pursued in the minority world are almost non-existent in South Asia. In the main uranium mine – Jaduguda – 200 trucks of ore pass through the village every day. Uranium tailings lie unprotected in front of the local school. The liquid waste from the mine is dumped in the tailings dams, which is then diverted into a channel, ending up in the Subernarekha River. In one of the villages, Chatikocha, 500 people live below the embankment of the tailings dam (Mahapatra, 2004). Saluka Himbram, the head of the village, talks of living next to the mine and the tailings dam:

> Abnormal births have become common. Half of the women have problems in delivery and miscarriages . . . We feel like vomiting when the wind, carrying fine dust from the pond, reaches us . . . The tailing pond must be causing the problem (quoted in Mahapatra, 2004).

The critical message here is that environmental security issues relating to uranium mining and nuclear energy will be magnified in the less affluent world. The cases of Australia and Germany may allow us to imagine a nuclear future for the Indian Ocean – to wrestle with potential waste issues – but the cold reality is, that the Indian Ocean, as a nuclear ocean, will exhibit levels of environmental degradation which can be scarcely imagined in the West. In simple terms, the lives of people are more expendable, in political terms, in the majority world, and this, in turn, has a devastating impact on their environments.

At the regional level, it is not just waste that is generated within the region, but also the ever-increasing waste which passes *through* the region. The Indian Ocean is often referred to as the Ocean of the South (Chaturvedi, 1998). As such, environmental practices are often excused on the basic reasoning that the lives of many people in the region are so environmentally-marginal, that a little bit more degradation wouldn't hurt. As a consequence, the Indian Ocean Region is currently being used as a key transport route for nuclear waste. The two biggest destinations for the processing of waste are Sellarfield, in England and Le Hague in France. Song writes:

> The first question in nuclear waste disposal is reprocessing, which can recover a significant quantity of useable material for nuclear reactors and reduce the amount of high-level waste that must be stored to about three per cent of the original spent fuel (Song, 2003, 8).

Waste reprocessing breaks down spent fuel, chemically dissolving it, with the plutonium separated from it. Apart from large, routine discharges of radioactivity from this high level and long-lived nuclear waste, this plutonium is now mixed with uranium (MOX) and re-used in conventional reactors. The Japanese are the biggest proponents of this process, and receive their shipments from both the British and French reprocessing plants. Unfortunately, this MOX fuel is often transported through the Indian Ocean on its way back to Japan.

It is impossible to know just how many of these shipments are being made, due to the high-levels of secrecy which surround them. What we do know, is that on 21 July 1999, two ships carrying weapon-usable plutonium, left Europe via the Cape of Good Hope, through the Indian Ocean to Japan. The task of publicising the shipments was left to transnational NGO, Greenpeace, who reported as follows:

> The two British flagged vessels, the Pacific Teal and the Pacific Pintail, left Barrow in Britain and Cherbourg in France carrying the first commercial shipment to Japan of mixed-oxide (MOX) reactor fuel, made from plutonium and uranium. An estimated 446 kilograms of plutonium is contained in the 40 nuclear fuel elements – enough fissile material to construct at least 60 nuclear bombs . . . The Cape of Good Hope has become the path of least resistance (Greenpeace, 1999).

Only Mauritius, acting alone, made public its opposition to the reprocessed fuel's transport, by refusing to admit the vessels into its Exclusive Economic Zone (African News Service, 7/3/99). Of course, this can only be a symbolic position, as the devastation which would occur if the vessel were sunk, or caught fire, would make the 200 km zone, as a zone of protection, look ludicrous. The point has already been made that, under existing liability agreements, there is some limited compensation under international conventions, "but no assurances exist whatsoever that the full costs of health, environmental and economic damages would be paid to victims in en route states" (Greenpeace, 1999).

Conclusions

Apart from dismissively being regarded as simply *en route states* for the international trade in waste reprocessing, countries of the IOR are also increasingly being looked at as possible repositories for international waste, as a nuclear dumping ground. If the only waste product of nuclear energy were high-levels of radioactivity for a half shelf life of 250,000 years (as is widely reported for high-level waste) then the Indian Ocean would already have been targeted. For, as is already well-documented by the environmental justice movement, it is the poorer states that accept the waste of wealthier societies, in exchange for money. And within those countries, it is the poorest communities who are forced to dwell alongside these repositories (Bullard, 1993).

At this juncture, the environmental security concerns of waste storage and transport become intermeshed once more with more traditional security concerns of nation-states and their defence. Because spent nuclear fuel can be reprocessed to produce both weapons of mass destruction – as well as energy – this placement of such waste dumps is highly politicised, with the United States increasingly involved in decisions as to who can and who cannot store and reprocess spent nuclear fuel. In fact, recently the US decided to store all its own spent fuel at Yucca Mountain by 2010 in Nevada, in large part, due to its fears that exported waste may find its way into the nuclear defence programmes of other states.

Of the Indian Ocean countries, as mentioned, Australia has been specifically targeted as a possible dump by transnational corporation, Pangea. In the late 1990s, Pangea placed the policy spotlight on Australia due to its apparent political and geomorphological stability. Though there is some merit to the geomorphological arguments, imagining that political stability could survive thousands of years is rather comical and nonsensical, at best, and downright dangerous, at worst. Currently, Pangea has backed off its Australian target, whilst the Australian Federal Government has taken over the running, this time advocating a low to medium density national waste repository in South Australia. As touched upon, both the state government of South Australia, the national opposition, and the people of South Australia remain vehemently opposed to such disposal on their doorstep.

In other countries in the IOR, both India and Pakistan store their own waste, and in the case of Iran, there appears to be some tension as to whether the waste from the Bushehr nuclear power plant will be transported to Russia, or will remain on home soil (Kerr, 2002, 29). Sections of the Iranian government want to store their own waste, whilst powerful elements in the international community – particularly the US – have made it clear that if the waste is not returned to Russia, then the nuclear programme within Iran should be terminated.

At the broader level of anti-nuclear movements across the globe, it is extremely interesting to see anti-nuclear movements emerging for the first time in the majority world. As aforesaid, these movements also share an active disdain for the build-up and utilisation of nuclear weapons within traditional security discourses.

The Indian Ocean is fast-becoming the nuclear ocean. Because of its status as the ocean of the south, the environmental security focus on nuclear power has not been sufficiently explored, as the possibilities of nuclear accidents whether in the mining of uranium, power generation, reprocessing, transport or storage are seen as risks which those less affluent are expected to take. With its increased nuclear profile, it is critical that environmental security concerns are included alongside the more traditional concerns of weapons proliferation and possible nuclear war.

Environmental security issues such as nuclear energy provide human security issues which cross the political boundaries of nation-states, gathering momentum in ecological, geopolitical regions such as the Indian Ocean Region. It is the immensity of these problems, paradoxically, which demand regional cooperation if they are to be successfully addressed. This cooperation, with great hope and conviction, may one day provide environmental security – human security - for all who dwell in the region.

References

Africa News Service, (1999) "Mauritius-Nuclear Mauritius Opposes Nuclear Waste Transfer," *Africa News Service,* August 2.

Barnett, J. and Dovers, S. (2001) "Environmental Security, Sustainability and Policy," *Pacifica Review*, vol. 13, no. 2, pp. 157–69.

Bidwai, P. and Vanaik, A. (2000) *New Nukes: India, Pakistan and Global Nuclear Disarmament*, Signal Books, UK.

Blowers, A. and Lowry, D. (1997) 'Nuclear Conflict in Germany: The Wider Context', *Environmental Politics*, vol. 6, no. 3: 148–155.

Bullard, R. (1993) *Dumping on Dixie*, Westview Press, Philadelphia.

Chaturvedi, S. (1998) "Common Security? Geopolitics, Development, South Asia and the Indian Ocean," *Third World Quarterly*, vol. 19, no. 4, pp. 701–24.

Clayton, G., Dalton, L., Foard, G., and Marshall, L (1986) "Chernobyl Fallout," *Chain Reaction*, 46, special section.

Dabelko, G.D. and Dabelko, D.D. (1995) "Environmental Security: Issues of Conflict and redefinition", Woodrow Wilson Environmental Change and Security Project Report, issue 1, pp. 3–12.

Die Tageszeitung (1995) 24–29 April quoted in Doyle, T: (2005) *Environmental Movements in Minority and Majority Worlds: A Global Perspective*, Rutgers University Press, New Brunswick, New York and London.

Doyle, T. (2000) *Green Power: the Environment Movement in Australia*, University of New South Wales Press, Sydney.

Doyle, T. (2004a) "Dam Disputes in Australia and India: Appreciating Differences in Struggles for Sustainable Development," in D. Gopal and D. Rumley (eds) *India and Australia: Issues and Opportunities*, Authors Press, Delhi, pp. 365–84.

Doyle, T. (2004b) "An Agenda for Environmental Security in the Indian Ocean Region," in D. Rumley and S. Chaturvedi (eds) *Geopolitical Orientations, Regionalism and Security in the Indian Ocean*, South Asian Publishers, New Delhi, pp. 154–71.

Doyle, T. (2005) *Environmental Movements in Minority and Majority Worlds: A Global Perspective*, Rutgers University Press, New Brunswick, New York and London.

Green, J. (2001) 'Germany: Greens Betray Anti-Nuclear Movement', *Green Left Weekly*, no. 444, April (online at www.geocities.com/ imgreen3/ ermangreens).

Greenpeace (1999) "Japanese Plans to Make Carribean the Toxic Throughway for Clandestine Shipments of Nuclear Waste and Plutonium", unpub. briefing, April.

Joppke, C. (1992) 'Explaining Cross-National Variations of Two Anti-Nuclear Movements: a Political Process Perspective', *Sociology*, vol. 26, no. 2: 311–321.

Kerr, P. (2002) "U.S. Reportedly Offers Russia Deal on Bushehr," *Arms Control Today*, November, p. 29.

Mahapatra, R. (2004) "Eyewitness: Radioactivity Doesn't Stop at the Mines in Jaduguda," *Down to Earth*, vol. 12, no. 23.

Mariotte, M. (1997) 'The Siege of Gorleben', *The Third Opinion: Australia's Independent Quarterly Energy Journal*, Autumn: 3–6.

Martin, T. (2001) 'Not Easy Being Green', *Europe*, February, issue. 403: 22–24.

Noonan, D. (Australian Conservation Foundation, Nuclear Campaigner) (2000) quoted in Doyle, T. (2005) *Environmental Movements in Minority and Majority Worlds: A Global Perspective*, Rutgers University Press, New Brunswick, New York and London.

Papadakis, E. (1998) *Historical Dictionary of the Green Movement*, Scarecrow Press, Maryland.

Rucht, D. and Roose, J. (1999) 'The German Environmental Movement at a Crossroads', *Environmental Politics*, Frank Cass Publishers, vol. 8, no. 1: 59 – 80.

Skullerud, J. (2001) Personal correspondence with the author, 12, 15 June.

Song, F. (2003) "Currently Indisposed: Managing Radioactive Waste," *Harvard International Review*, Summer, pp. 8–9.

Summy, R. and Saunders, M. (1986) *A History of the Peace Movement in Australia*, published by University of New England, Armidale, Australia.

CHAPTER 13

Energy Security and Earth Rights in Thailand and Burma (Myanmar)[1]

Adam Simpson[2]

> We urge freedom-loving governments everywhere to impose sanctions on this illegitimate regime. They worked for us in South Africa. If applied conscientiously, they will work in Burma too.[3]
>
> Archbishop Desmond Tutu

Introduction

Human rights and environmental protection and large-scale energy projects are uncomfortable bedfellows. 'Energy security' is often cited as an excuse to pursue the latter, even at the expense of the former. It is the nature of the present geopolitical situation that many such scenarios occur throughout the world – brutal regimes supported by the foreign exchange received, and the political capital earned, by undertaking large-scale energy projects with Western multinational corporations. This situation currently exists in Burma as the forty year-old military junta struggles to survive, both economically and politically.

Although the Thai government and the Burmese military ruling junta, the State Peace & Development Council (SPDC), have an uneasy relationship, they have managed to set aside their differences on a number of occasions to enter significant energy deals – often citing energy security as their raison d'être – with multinationals that have had a detrimental impact on human rights and environmental integrity on both sides of the border.

One of the most important of these deals was the construction of the Yadana gas pipeline that has resulted in over a decade, and continuing, gross and systematic human rights and environmental abuses on the Burmese side with the full knowledge of the Thai government and the two multinationals involved – Unocal of the US and Total of France.

This project has met with local and international resistance since its inception and is now the subject of a landmark court case in the US. But for the long-suffering people of Burma, and particularly the ethnic minorities in the path of the pipeline, it has been an unwelcome crutch to the military regime and an endless source of pain and suffering.

The concept of *earth rights*, as the nexus between human rights and environmental protection is a useful, and developing, concept for the analysis of such projects. This chapter examines the issues and activism surrounding the building of this pipeline, and other energy projects that may have similar destructive impacts in the future, through the dual conceptual prisms of *energy security* and *earth rights*.

Energy Security

"Energy security" is a nebulous concept, but to most contemporary governments it relates to the state's ability to produce enough electricity for unlimited Western-style development and enough oil to power increasing numbers of automobiles, buses, trucks and factories. In the context of Thailand and Burma, Thailand has used significant energy resources for its industrialisation over many decades, while Burma, until recently, has used relatively little.

Both Burma and Thailand possess significant energy reserves, with Burma announcing in early 2004 the discovery of another huge natural gas deposit in the Bay of Bengal, possibly twice the size of the recoverable gas of the Yadana field (Myo Lwin, 2004, January 19-25). The relatively small needs of Burmese industry at present usually results in the export of its energy across the border to Thailand, although there are plans for pipelines to other countries such as India (Puri, 2004, January 29).

In relation to electricity generation, on which this chapter primarily concentrates, the focus in both countries has been on big dam construction and diesel or gas-fuelled power stations, strategies which favour large construction and oil or gas companies. Both these processes have environmental and human costs that are rarely considered when calculating the cost of the energy produced, resulting in favourable comparisons with more 'ethical' forms of electricity generation, such as solar power (Laird, 2000, 341). In addition, the politics of money and patronage in Thailand have often resulted in poor investment decisions being made on the basis of political favours and personal gain rather than on the basis of national benefit.[4]

The politics attached to electricity-generating technologies, as with all technologies, should also not be overlooked. Technologies such as large hydroelectric schemes and fossil fuel electricity generation require *centralised control* and extremely large financial investments, providing governments and businesses with mouth-watering opportunities for corruption and enhanced centralisation of control and political power. Solar and wind power generation reduces dependence on government and large corporations and can lead to similarly subversive political aspirations.

Given the large energy reserves of the two countries, the main focus has been on extraction, delivery and conversion into sufficient electricity to ensure unrestricted industrial development. Thailand has been particularly successful at creating an excessive electricity supply with a projected reserve margin of over 50 per cent (Giannini, Redford, Apple, Greer and Simons, 2000, 167). Consequently, the aim of Thai Prime Minister Thaksin Shinawatra is for Thailand to be a regional energy-exporting hub facilitating the *energy security* of all Asia (Balan Moses, 2003, July 9).

The Burmese SPDC has similar aspirations with the former Prime Minister and dominant powerbroker of the Burmese military regime, General Than Shwe, suggesting Burma could become a "reliable source for the region's . . . energy security without failing in its responsibility to contribute towards regional peace, security and prosperity" – a responsibility, some would argue, it has yet to fulfil (Myanmar leader returns from ASEAN Summit Meeting, 1998, December 17).

In addition to the projects discussed in this chapter there are many similar large-scale energy projects planned for the northern and western Indian Ocean Region and as Aparajita Biswas notes in Chapter 5 in this volume, of all energy sources the demand for natural gas is likely to grow the fastest. Gas pipeline projects from Burma to India via Bangladesh and Iran to India via Pakistan are becoming ever more likely (Devraj, 2004, June 8). Most states in the region are therefore likely to be engaged in such projects in the name of 'energy security' in the near future.

The concept of 'energy security', therefore, has significant political connotations attached to it, and, while a comprehensive examination of the concept is beyond the scope of this chapter, the following analysis should be digested keeping in mind the large-scale industrial development paradigm associated with the common use of the term.

Earth Rights

The dawning realisation throughout the industrialised world of linkages between human rights and environmental protection, previously considered separate areas, has seen the rise of the concept of earth rights among disparate groups such as Amnesty International-USA and the Sierra Club.[5] It is, however, the *nexus* between the two issues that corresponds to the increasingly accepted earth rights moniker and therefore there are some aspects of the issues, such as pure wilderness conservation in the US, that would not generally be included under the earth rights banner (Redford, 2004, pers. comm.).

According to Greer and Giannini (1999, 20):

[e]arth rights are those rights that demonstrate the connection between human well-being and a sound environment, and include the right to a healthy environment, the right to speak out and act to protect the environment, and the right to participate in development decisions.[6]

The environment group *EarthRights International* was founded in 1995 by two American lawyers, Katharine Redford and

Tyler Gianinni, and a Karen exile from Burma, Ka Hsaw Wa,[7] who all held similar beliefs on this issue. Ka Hsaw Wa, while originally a proponent of armed rebellion against the Burmese military for the murder of his student friends in the 1988 massacres, came to abhor all violence and turned to nonviolent methods. He started gathering evidence of human rights abuses from interviews with predominantly ethnic Burmese exiles and found that a common thread in their horrific stories was the destruction of their environment by the military. Through this process, Ka Hsaw Wa came to realise the vital interconnectedness of the two issues of human rights and environmental protection (Ka Hsaw Wa, 2004, pers. comm.).

Giannini and Redford came from different backgrounds but both similarly understood the symbiotic nature of these two issues. They then joined with Ka Hsaw Wa to form *EarthRights International* whose activities were to focus on the nexus of environmental protection and human rights through the rule of law and the power of people (Redford, 2004, pers. comm.). EarthRights International has been intimately involved with the investigation of earth rights violations along the Yadana gas pipeline. Ironically, the atrocious violations of earth rights that have occurred in connection with the pipeline have been a stimulus for the development of the earth rights concept and for the cooperation between previously irreconcilable groups.

In future, connections and linkages between the concepts of earth rights and 'environmental security' (Doyle, 2004) are likely to provide fertile opportunities for delivering improvements both to the ecological and social well being of communities in the Indian Ocean Region. As the convention in this chapter employs the concept of earth rights, when human rights or environmental issues are mentioned it is important to connect any environmental degradation with an erosion of human rights and vice versa.

Yadana Gas Pipeline

The Yadana Gas Project was formally initiated in July 1992 when the French corporation, Total, signed a contract with the Burmese company, Myanmar Oil and Gas Enterprise (MOGE). The purpose of the contract was the "evaluation, development and production" of gas in the field of Yadana located in the Andaman Sea, Martaban Gulf, 70 kilometres from the southern coast of Burma.

By February 1995, Unocal and PTT-EP had joined the partnership with the following financial investments in the project:

- MOGE – Myanmar Oil & Gas Enterprise (Oil co. of the Burmese State) – 15%
- Unocal – Union Oil of California – 28.26%
- PTT-EP – Petroleum Authority of Thailand Exploration and Production Public Co Ltd – 25.5%
- Total – French Consortium – 31.24% (US$700m in 1996)

The pipeline is approximately 700 km long: 409 km of which are in Burmese territory – 346 km off-shore and 63 km on-shore across Tenasserim Division in the south east of the country – and 260 km of which are on-shore in Thai territory. The pipeline terminates at Ratchaburi, where the Electricity Generation Authority of Thailand (Egat) has built a power plant that converts the gas to electricity.

In examining the Yadana gas pipeline it should be noted that the Yetagun gas pipeline project, also drawing gas from the Gulf of Martaban and developed later by UK company Premier Oil, Petronas of Malaysia, Nippon Oil, PTT-EP and MOGE, has been involved in the same human rights and environmental abuses as the Yadana pipeline project. The pipeline follows a separate, but parallel, path to the Yadana pipeline across Burma until the Thai border where the pipelines join and a single pipeline continues to Ratchaburi (Figure 13.1).

Thai Energy Requirements

At the time of the Yadana project's inception future gas sales to Thailand were estimated at between 200 to 400 million USD annually, amounting to about half of all Burma's export revenue for 1994-95 (Laroche, 1996). Energy consultants Wood and Mackenzie of the United Kingdom estimated that Burma will earn a nominal US$2.87 billion over the expected life of the field to 2030 – this amounts to a significant amount of foreign currency for a country whose 1998 foreign exchange reserves amounted to between $150 million and $350 million (Prospects poor for Burmese gas, 1998).

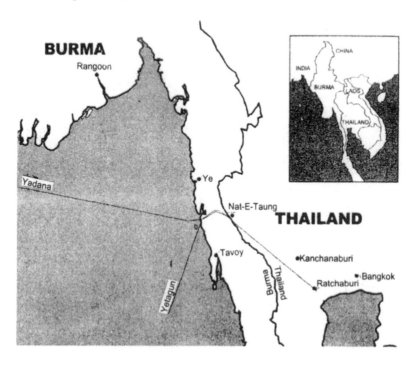

Figure 13.1. The Yadana and Yetagun pipelines.
Source: Total Denial Continues (Giannini, Redford, Apple, Greer, and Simons, 2000, 12)

Gas was contracted to be delivered to Ratchaburi by July 1998, but neither the Thai pipeline, being constructed by PTT contractors Tasco Mannesman, nor the Ratchaburi plant, the responsibility of Egat, were completed on time. The contract the PTT signed with the Yadana consortium committed it to taking, or paying for, at least 65 mcfd (million cubic feet per day) on completion of the pipeline. According to its own estimates, the PTT paid US$81.57 million to the consortium because it failed to take the full contracted natural gas supply within five months of the agreed starting date for gas purchases – the cost of which will be passed on to the Thai people (Egat pushes for 13% increase in base tariff, 1998; Pipe dreams become nightmare for Thailand, 1998).

The Ratchaburi plant began to take a limited volume of gas from Yadana in December 1999 after a delay of 18 months. In January 2000, PTT temporarily reduced the charge for Burmese gas to 17 baht per one million BTU (British Thermal Units), making it 10 per cent cheaper than fuel oil and domestic gas for power generation as PTT charged 19 baht per one million BTU for gas from the Gulf of Thailand (PTT cuts transport cost of Burmese gas, 2000).

In the Thai Government's search for 'energy security', the state itself has suffered huge losses over the project having to maintain a high electricity reserve of 5,000 megawatts that even in 2000 was estimated to have cost 60,000 million baht (US$1.4 bn) (PTT faces lawsuit over Yadana project, 2000). From August 1998 to December 2001, the Yadana and Yetagun projects alone, where the contract is take-or-pay for thirty years, cost Thailand US$948 million when the Ratchaburi plant only became fully operational in April 2002 (Janchitfah, 2002).

Projections by Egat in 1999 suggested that, in 2003, Thailand's reserve margin – that is, the amount of generation capacity beyond actual need – would be 52% (Giannini et al., 2000, 167). This massive surplus capacity is incredibly wasteful and results in the Thai government paying significant amounts of interest on electricity generation capacity that it neither wants nor will need in the foreseeable future.

The huge reduction in power demand throughout 1998 following the Asian financial crisis finally forced Egat in September of that year to reverse its plans to increase the electrical supply by 6,300 MW over two years and to *reduce* supplies by 6,000 MW (Power Play, 1998). This estimate did not prevent Egat from borrowing US$240 million from the Import-Export Bank of Japan later that month, over half of which was designated for the Ratchaburi power plant (Egat signs B9.3 billion loan deal, 1998).

The continuing energy glut was demonstrated in December 2002, when a Thai Senate Committee on Public Participation recommended the year-round opening of the sluice gates of Pak Moon dam, restoring the ecological balance to the Moon river, as the hydro-electricity was not needed for the next five years (Senators: Open the dam gates, 2002).

The construction of the Yadana pipeline to provide this excess capacity has had numerous detrimental impacts on local communities and their environments.

Environmental Impacts in Thailand

There are a number of issues relating to the Environmental Impact Assessment (EIA) of the project that require investigation; a brief survey of some aspects follows. EIAs are relatively underdeveloped in most parts of Asia as its more recent economic development compared to the West has only recently produced the more obvious forms of environmental degradation.

The Thai Government has, since the late 1980s, made some efforts to minimise environmental impacts of large projects by passing environmental legislation to control development. One aspect was an attempt, in 1992, to make EIAs mandatory prior to governmental approval for projects such as Yadana. Unfortunately, in this case, government assent was given in 1995 while the EIA was only completed two years later in June 1997. In addition to the illegal nature of this process, the Environmental Impact Statement (EIS) was produced only in English, which resulted in the large majority of the residents in Kanchanaburi Province who were to be affected unable to access it. The language used in EIAs is often complex enough for native speakers, and being in English meant that even activists who understood some English had little chance of fully understanding the document.

There were many inconsistencies between the text of the EIS and the reality of the project. The EIS assured that only 'small' trees would be cut down to make way for the pipeline, but during protests, groups of up to six protesters linked hand in hand measured the circumference of some marked trees, at least six metres, which were later turned into stumps. While promising minimal disturbance to the environment, in reality, the construction destroyed 'large tracts of pristine, ecologically vulnerable forests' (Killing what is its job to protect, 1998). The EIS also assured that the pipeline would not pass through elephant habitat while the Thai World Wildlife Fund estimated there were 40-50 elephants in the area. During construction of the pipeline, at least six elephants were trapped in the pipeline trench at different times. At least three of these elephants perished (Phinan Chotirosseranee, 1998, pers. comm.).

The pipeline route passed through several protected areas, including a national park and conservation forest areas. In an apparent attempt to appear magnanimous by taking the least-worst alternative, the EIS briefly mentioned two other possible routes for the pipeline but both of these passed through Class 1A watershed areas and longer stretches of conservation areas than the final route when more sensible routes with less economic and ecological damage were available (Giannini et al., 2000, 142-3).

The technical specifications of the EIS stated that the pipe would be 17.1mm thick. This statement was proved to be inaccurate by measurements in the field by the Kanchanaburi Environment Group showing that the pipeline was, at most, 16.1mm thick (Phinan Chotirosseranee, 1998, pers. comm.).

It was also claimed that reforestation would be undertaken using species from the original forest. Two breaches of this commitment were reforesting using mango trees and grasses. For the former, the Forestry Department may have had one eye on future revenue opportunities, and for the latter it was much easier and cheaper than planting trees. Grasslands occurring in the middle of the forest increase the chances of fires in the area during the dry season as the forest usually remained moist enough to prevent fires while grassed areas dry out quickly. The prospect of a fire occurring over a buried gas pipe that had already shown several leaks concerned many of the local residents (Pressure builds over Yadana pipeline leak, 1998).

Following these breaches of the EIA legislation and economic loss from the project, the local Kanchanaburi Environment Group vowed to take legal action for economic and environmental damages (PTT faces lawsuit over Yadana project, 2000; Phinan Chotirosseranee, 1998, pers. comm.).

Engagement with Burma

Many groups, both inside and outside Thailand, have been concerned with the linking of the Thai Government with the Burmese military junta, and therefore its repression, through the Yadana project. Essentially, the same Burmese military regime has been in power since General Ne Win grasped power in a military coup in 1962. In 1988, the regime brutally crushed massive pro-democracy demonstrations with at least 3,000 protesters killed. Pro-democracy leader, Aung San Suu Kyi, was placed under house arrest in 1989, along with other leaders of the National League for Democracy (NLD) and the results of the 1990 election, where the NLD won 392 of the 485 seats, have been ignored by the military junta ever since.

Following advice from a Washington-based public relations company, the junta changed its name from SLORC to the State Peace and Development Council (SPDC) in 1997, but, in essence, it remains the same to the present day. Ne Win died in 2002, but his protégé, General Than Shwe, remains influential within the SPDC despite the appointment of General Khin Nyunt as Prime Minister in September 2003.

Many Western companies have shied away from being associated with the regime and, in 1997, the US government applied sanctions banning American companies from any new investment in Burma.[8] This has not, however, affected Unocal, as the Yadana project preceded 1997 but many institutional investors have sold Unocal and Total stock due to their involvement with the Burmese military regime.[9] The *Burmese Freedom and Democracy Act 2003*, passed unanimously by Congress and signed into law on July 28th 2003, bans all investment in Burma, but the impact of this on Unocal is uncertain as financial institutions are finding ways around the ban (Bush, 2003; Mathiason, 2004, January 18). The UK company, Premier Oil, after years of dismissing human rights abuse claims by NGOs, agreed in September 2002 to sell its interests in the Yetagun gas pipeline, which has also been the focus of human rights campaigners (Berton, 2002; Macalister, 2002; 2003).

The allegations of human rights violations in Burma, that have occurred for decades and continue to the present day, are not seriously questioned by outsiders without a vested interest in the country. The repression that the Burmese military, the *Tatmadaw*, delivers daily to the ethnic minority groups and their environments occurs all along the border with Thailand but has been particularly concentrated along the route of the Yadana pipeline in Tenasserim Division. Indeed, a 1992 assessment by the consulting firm, Control Risk Group, paid for by Unocal, concluded that the Burmese military "habitually makes use of forced labour to construct roads", adding that the *Tatmadaw* was ordering whole villages to relocate, dislocating the predominandy animist ethnic minorities from their environment and spiritual homes, the goal being to sever any ties the villagers had with Karen and Mon rebels.[10] Additional environmental attacks on these minorities included the destruction of villages, crops and forests.

The report also noted that "there are credible reports of military attacks on civilians in the region [and] the local community is already terrorised" (Girion, 2003, June 15). For Unocal to engage in a major construction project and to later argue that it had no responsibility for the actions of the *Tatmadaw* on the project, as it does in *Doe v Unocal Corp* (see *Alien Tort Claims Act 1789* below), is farcical, especially given the signed testimony of Unocal's former President, John Imle, admitting that "some porters were conscripted" (Giannini, 1999, 15).

In April 2003, two months before a crackdown on Aung San Suu Kyi and the National League for Democracy (NLD), the UN Commission on Human Rights expressed its grave concern at the situation in Burma due to, amongst other issues:

> Extrajudicial killings; reports of rape and other forms of sexual violence persistently carried out by members of the armed forces; continuing use of torture; renewed instances of political arrests and continuing detentions, including of prisoners whose sentences have expired; forced relocation; destruction of livelihoods and confiscations of land by the armed forces; forced labour, including child labour; trafficking in persons; denial of freedom of assembly, association, expression and movement; discrimination and persecution on the basis of religious or ethnic background; wide disrespect for the rule of law and lack of independence of the .judiciary; unsatisfactory conditions of detention; systematic use of child soldiers; and violations of the rights to an adequate standard of living, such as food, and to medical care and to education.
>
> (United Nations Commission on Human Rights, 2003)

These human rights abuses have been exhaustively documented by human rights groups such as EarthRights International, other

NGOs and government bodies (Bureau of International Labor Affairs, 1998; Giannini, et al., 2000; Greer, 1998; Htoo, Maung, Kher, Myo, MacLean, Imamaura, & Giannini, 2002; ILO, 1998; Laroche, 1996).

Given the lack of freedom of movement, freedom of expression and the generally repressive culture that dominates throughout the country, much of this research has been undertaken covertly. The US Government, however, generally agrees with these assessments and estimated that at least five million cases of forced labour had occurred from 1988 through 1996 (Bureau of International Labor Affairs, 1998).

Forced Labour

Although these human rights abuses occur throughout Burma, some of the worst abuses have been carried out near the route of the Yadana gas pipeline where villagers have been used as slave labour on the pipeline project or as porters carrying supplies or ammunition for soldiers patrolling the pipeline route. Often villagers do not survive the ordeal under the extreme conditions but face summary execution if they do not cooperate (Giannini et al., 2000). Forced labour in this area continues to be linked to two battalions – the Light Infantry Battalions (LIBs) 273 and 282 – that were created to provide security for the Yadana and Yetagun gas pipelines in 1995 and 1996. These battalions are known by local villagers as 'Total' battalions and have been continuously linked with forced labour and violence from their inception (Myo, Imamura, Foley, Robinson, Maung, Htoo and Giannini, 2003b, 5).

In addition to forced labour and portering for the soldiers, villagers are also required to work as human mine sweepers. The military has automated mine sweepers but villagers are often used instead to discourage ethnic insurgents from laying land mines (Myo et al., 2003a, 21-3).

Systematic Rape

Since adoption of the Rome Statute of the International Criminal Court in 1998, systematic rape of a civilian population has been considered a crime against humanity (1998, Article 7 (1)(g)). In its pursuit of 'energy security', the Burmese military appears to have undertaken such a strategy against ethnic minorities as documented in the following reports.

In April 2003, Refugees International published a report, *No Safe Place: Burma's Army and the Rape of Ethnic Women* (Apple, 2003), documenting 43 rapes of ethnic women by *Tatmadaw* soldiers. Nine of these rapes resulted in death, 12 were gang rapes and 15 of the perpetrators were officers of the *Tatmadaw* who were known to the victims (Apple, 2003, 70). These documented rapes occurred mostly along the Thai-Burmese border with several occurring in the Yadana pipeline region. Anecdotal evidence of hundreds of cases was gathered during a month of interviews but not all could be pursued (Apple, 2003, 57). This report is supported by another issued in the same year by the Shan Human Rights Foundation (SHRF) and the Shan Women's Action Network (SWAN) entitled *Licence to Rape* (2002) that documents 173 incidents of rape and other forms of sexual violence by the *Tatmadaw* within Shan State in north east Burma.

In April of the following year, the Karen Women's Organisation, formed in 1949, released a further report (2004) documenting 125 cases of rape by the *Tatmadaw* against Karen women, predominantly in the Thai-Burmese border region near the Yadana gas pipeline. This report documents cases where, for example, a column commander raped a woman then ordered his subordinates, on pain of death, to do the same and an eight months pregnant woman was raped by eight soldiers. These were not isolated incidents.

These reports demonstrate that sexual violence is not simply a by-product of *Tatmadaw* operations, such as clearing the corridor for the Yadana pipeline; rather, that it is a systematic political strategy for dealing with the ethnic 'insurgents' that resist the military's policies of repression.

The Alien Tort Claims Act 1789

Despite the plethora of human rights abuses and environmental degradation that have resulted from the gas pipeline projects, there is one aspect to the Yadana case that could dramatically change the rules for conducting international business across the globe. The landmark case, *Doe v. Unocal Corp.*[11] *(1997)* has caused grave unease in boardrooms across the United States and given a semblance of hope to victims of human rights violations involving American multinationals, wherever these violations have occurred.

In 1997, a US federal district court in Los Angeles agreed to hear the case and concluded that corporations and their executive officers could be held legally responsible for violations of human rights norms in foreign countries and that United States courts have the ultimate authority to adjudicate such claims (http://www.earthrights.org/unocal/index.shtml).

The obscure legislation this case is based upon is the *Alien Tort Claims Act* 1789 (ATCA) that grants original jurisdiction to any civil action claimed by an alien for a tort committed in violation of 'the laws of nations or a treaty ratified by the US' (Christmann, 2003, 209). The most widely accepted international law norms are also referred to as 'jus cogens', and include genocide, torture, systematic racial discrimination and slavery.

The ACTA was probably initially enacted to allow US courts to hear piracy cases but had lain virtually dormant for almost two centuries until *Filitarga v. Pena-Irala* (1980), when the Paraguayan plaintiff successfully sued for civil damages against a Paraguayan police officer living in New York who had tortured his son to death in Paraguay. Labelling torturers 'enemies of all mankind', the Court of Appeals for the Second Circuit allowed the case to proceed under the ACTA (Slaughter and Bosco, n.d.).

The present Unocal case is different to the Filitarga case as the defendant is a body corporate rather than an individual. The plaintiffs in the Unocal case are fifteen Burmese villagers who claim that they were subjected to forced labour, rape, and torture during the construction of a gas pipeline through their country. Soldiers allegedly committed these abuses while providing security and other services for the pipeline project, making Unocal *vicariously liable* (Mariner, 2003; http://www.earthrights.org/unocal/index.shtml). Following various dismissals and appeals, on 17 June 2003, eleven judges of the Ninth US Circuit Court of Appeals started to hear, *en banc*, arguments over whether the Burmese plaintiffs can sue Unocal Corp. in a U.S. federal court for the alleged brutalities (Tam, 2003).[12]

The implications of this case for American multinationals, particularly energy companies that often do business with repressive regimes, is significant. New York's federal district court has cases pending against Fujitsu, Unisys, Citigroup, Credit Suisse, IBM, Deutsche Bank, Dresdner Bank, ExxonMobil, Ford and GM. An existing case focuses on Royal Dutch/Shell's connection to human rights abuses in Nigeria. In Houston, an action was filed against ExxonMobil and Shell (The Alien Problem, 2003).

The cases are causing companies to examine more closely the human rights record of the governments they do business with and "causing companies to run away from situations like Burma" (Markels, 2003). Companies have already pulled back from countries where governments have been accused of human rights abuses, including Nigeria, Indonesia, and Sudan as well as Burma (Markels, 2003).

It is therefore not surprising that the Bush administration, which has an extremely close relationship to big business in general and the energy sector in particular, has attempted to stifle both the Unocal case and the ATCA in the form of a 'friend of the court' brief from Attorney General John Ashcroft filed on 8 May 2003 (Mariner, 2003). Arguing that interpreting the ACTA in favour of the plaintiffs would undermine the Administration's 'war on terrror', Ashcroft is, in reality, undermining the prospect of international justice and protecting corporate interests from victims of their profiteering in countries run by authoritarian regimes (Tyranny and terrorism, 2003).

This case will, in all likelihood, continue for several years, but its very existence is already having positive benefits for those who would otherwise suffer due to multinationals propping up despotic and repressive regimes.

Earth Rights in Thailand

While examining earth rights violations in Thailand and Burma it is easy to focus on the Burmese side simply because the violations are of such an horrific and widespread nature. The differences between the tactics of the Burmese military and the Thai government, are, however, narrowing. The Thai government undertook, between February and May 2003, over 2,500 extrajudicial killings related to the crackdown on alleged drug traffickers (Bhumiprabhas, 2003). This led the UN Special Envoy for Human Rights, Hina Jilani, to describe the situation facing

Thai civil society movements as encouraging a 'climate of fear' (Chimprabha, 2003). Ms. Jilani also expressed her concern at a security advisory from the Prime Minister, Thaksin Shinawatra, to the Foreign Ministry to try and stop foreign funding from reaching NGOs in Thailand (Chimprabha, 2003). Many NGOs have been raided by police and shut down. EarthRights International has avoided recent attention but still maintains a secluded and unsignposted suburban office where only trusted individuals are taken.

The greater openness and press freedom of Thai society is, of course, commendable relative to the situation in Burma, but human rights abuses occur and are occurring on a more frequent basis.[13] Certainly, having freedom of the press allows voices to document and oppose these actions, but as in much of the rest of the world, the ubiquitous phantom of 'national security' is being increasingly used to quell dissent (Ekachai, 2003, June 26; Yoon, 2003).

One of the most vocal journalists has been Sanitsuda Ekachai, now Assistant Editor at the respected *Bangkok Post*, who has often spoken out about government repression. While the press freedom enjoyed by Thais is only dreamt about in Burma, the limited ability to hold the government to account has only had a minor impact on the government's actions (Ekachai, 2000, June 7; 2003, January 16; 2003, June 26).

Prime Minister Thaksin's government has blatantly violated the country's Constitution as well as a number of international treaties on human rights. At the 59th Session of the UN Commission on Human Rights early in 2003, the issue was raised during an oral intervention made by *Pax Romana*, a Geneva-based human rights organisation, which accused the government of having failed to take effective measures to ensure the right to life of its people while at the same time creating a 'culture of impunity' (Bhumiprabhas, 2003). Such an oral intervention against Thailand is unprecedented in recent times while it is an annual occurrence for Burma. Thailand usually has a relatively sound regional record of human rights protection so it is particularly disconcerting to note this deterioration in the situation.

Other Energy Projects

The disturbing authoritarian trend in Thailand has been witnessed across society but is at its most visible on various energy projects across the country. There are several other energy projects being undertaken in both Thailand and Burma that are arousing local protests and alleged human rights abuses. Two examples are examined briefly below – the Thai-Malaysian pipeline in Thailand and the Tasang Dam in Burma.

Thai-Malaysian Pipeline – Thailand

Another gas pipeline project that has attracted considerable attention by local NGOs is the Thai-Malaysian pipeline heading through Songkhla Province in the south of Thailand into northern Malaysia. The Petroleum Authority of Thailand (PTT) and Malaysia's Petronas are joint operators of the proposed 30 billion baht (US$700 million) gas pipeline project, although American company, Amerada Hess Corporation, and British company, BP Amaco Plc, have been involved in surveys and production. The local NGOs who oppose the pipeline on environmental and economic grounds have been active since 1998 and have received attention and support from academics across the country (Kongrut, 2003, June 26).

The movement against the pipeline is now the largest civic group in the country and the government's increasing disregard for human rights has been showcased with a number of arrests of protesters on dubious charges and an increased security presence in the area (Ekachai, 2003, June 26). In early 2003, 400 police were deployed to guard the site of the proposed gas separation plant (Pipob Udomittipong, 2003, pers. comm.).

Following a brutal dismantling of a peaceful protest in Hat Yai in December 2002, the protesters have been gathering on Lan Hoy Saeb, a nearby beach. During the Hat Yai protest, twelve NGO members were arrested and, in tactics familiar to all social activists, the raiding police party removed their name and position badges from their uniforms (Janchitfah, 2003). In response to the police presence, a convoy of 300 local children on their bicycles asked them to leave the area and distributed leaflets and flowers to the police at the gas separation plant site (Kongrut, 2003, June 30). The UN Special Envoy for Human Rights, Hina Jilani, visited the area in May 2003 and deplored the arrest of local activists and called on Thaksin to improve human rights and public participation in the project (Chimprabha, 2003).

Although systematic human rights violations of the widespread and brutal nature of the Burmese regime are not expected on this project, the Thai Government's increasingly arrogant and militaristic activities, supported by Western energy companies such as Amerada Hess and BP Amaco, do point to significant and continuing human rights and environmental abuses on this project if it proceeds.

Tasang Dam – Burma

A proposed dam on the Salween River in Shan State, Burma, is another energy project with massive potential environmental and human rights implications. The Salween is the longest free-flowing river in Southeast Asia and the Greater Mekong Subregion (GMS) Power Co. Ltd. of Thailand and the Myanmar Economic Corporation of Burma are planning to block the flow with the tallest dam in Southeast Asia, at 188m, at Tasang in Shan State. The cost of the dam is estimated to be US$3 billion and the flood area caused by the dam would cover at least 640 square kilometres (http://www.earthrights.org/tasang/facts.shtml). The dam would produce up to 5,000 megawatts and an estimated storage capacity of 120,000 cubic metres compared with 6,000 cubic metres in the Bhumibol dam, the biggest in Thailand (Praiwan, 2002).

The Electricity Generation Authority of Thailand (Egat) is keen to invest in the project as it will avoid the protesting by environmentalists such projects have attracted in Thailand and it will not require an EIA, as it would in Thailand. Egat has agreed to develop, finance and operate the project for the duration of a concession granted by the Burmese government. At the end of the concession, Egat and Burma would have an equal stake in the project (Praiwan, 2002).

The project has been in the pre-construction planning stage for several years and has been waiting for finance and a buyer of the electricity. With Egat announcing it will now do both, the project should proceed in the absence of international pressure to abandon it. The Burmese military junta views the project as a way of securing a large proportion of its future energy needs and Egat, despite continually overestimating Thai energy requirements by vast amounts, is keen to develop energy sources in Burma due to political and environmental factors that attempt to restrain rampant development in Thailand.

In the 'developed' world, the era of large dams is ending. The Rasi Salai Declaration, endorsed at the Second International Meeting of Dam Affected People and Their Allies in December 2003 at Rasi Salai in Si Saket Province, Thailand, demonstrates the ongoing international struggle, particularly in the 'developing' world, to preserve indigenous and local peoples' land and culture against destructive large-scale dam projects (http://www. earthrights.org/news/riversforlifedeclaration.shtml) The proposed Tasang Dam fails the five criteria points of the World Commission of Dams Report (2000, 262-3) and is therefore neither in the interests of the people of Burma nor Thailand.

The Depayin Massacre

In May and June 2003, the Burmese military regime cracked down on pro-democracy activists across the country. While not directly related to energy projects, it is instructive to briefly investigate the crackdown as it is indicative of a more general policy of repression by the SPDC that becomes particularly severe in the vicinity of large-scale energy projects.

As noted earlier, pro-democracy leader and Nobel Peace Laureate, Aung San Suu Kyi, has been under house arrest, or some form or restricted movement, for most of the time since she was first placed in detention in 1989. As the most visible National League for Democracy (NLD) leader, both nationally and internationally, the winds of policy change within Burma have most often been aligned with her treatment. After being released from house arrest in May 2002, there were hopes of a mellowing of the regime's hard line policies, but, on Friday, 30 May 2003, all this changed near Depayin in northern Burma as Suu Kyi and her NLD convoy were attacked by "thousands of USDA [Union Solidarity and Development Association] members, plain clothes members of the security forces and freed convicts armed with sticks, machetes and some firearms who had been transported into the area with the help of local military commanders" (Baker, 2003, June 7-8).

During the massacre, at least 10 people died, 108 people 'disappeared', and 134 were sentenced or arrested (Ad hoc Commission on Depayin Massacre, 2003, 48). Both Aung San Suu Kyi and the deputy NLD chairman, U Tin Oo, were taken onto 'protective custody' under article 10a of the 1975 State Protection Law.[14] These accounts were reported by Burmese exile groups, which were then supported by the United States diplomats in Rangoon (Baker, 2003, June 7-8).

The conservative Thai government of Thaksin Shinawatra continued its policy of non-intervention suggesting it was an internal matter (Kaen, 2003). There were, however, elements of dissent within Parliament as the Senate Foreign Affairs Panel called on the government to review its relationship with the military junta (Treerutkuarkul, 2003). Panel chairman, Kraisak Choonhaven, argued that the arrest of Suu Kyi was "no longer an internal matter of the Burmese government", that Suu Kyi and her party members be released unconditionally and that the government should abandon its long-held stance of non-intervention (Treerutkuarkul, 2003).

The UN special envoy to Burma, Razali Ismail, was urged by human rights groups to demand a meeting with Suu Kyi as a precondition for his planned visit to Rangoon (Protestors urge condemnation, 2003, June 5). The next day he announced that this was indeed his intention (Razali puts general in a fix, 2003, June 6). He managed to secure an interview with Suu Kyi at that time but in the following 12 months, having undergone surgery and been relocated to her home – under house arrest again – she had only extremely limited access to outsiders. Suu Kyi was detained under article 10a of Burma's 1975 State Protection Law that allows de'ention without access to family or lawyers for up to 180 days at a time, up to a total of five years, suggesting Suu Kyi's internment is indefinite (Woods, 2003).

The decision by the SPDC to undertake the National Constitutional Convention in June 2004 without releasing Suu Kyi and NLD Chairman U Tin Oo, which would have been the ideal international public relations moment for reconciliation, indicates the unlikelihood of real progress (Baker, 2004, June 5).

Despite overtures to the KNU, the crackdown suggests that the Burmese military rulers are a long way from easing their grip on power.[15] While international investment continues, and international protests from governments such as Australia are muted, arbitrary detention of democratically-elected representatives of the people will continue unabated.

To Engage or Not to Engage

The arguments for engaging with the Burmese military regime are not convincing. The regional group most likely to influence Burmese policies, the 10-member Association of Southeast Asian Nations (ASEAN), has steadfastly maintained its policy of 'non-intervention in the internal affairs of a member state' and, until recently, has largely ignored the repressive activities of the military junta. At the Annual Conference of ASEAN in Phnom Penh in June 2003, following the attack and detention of Aung San Suu Kyi and other NLD leaders, the final communiqué did refer to Suu Kyi by name for the first time and acknowledged 'the incident' of her arrest, but then went on and congratulated the regime on its efforts to promote 'peace and development' (Baker, 2003, June 21-22). Such international support does nothing to influence the regime to alter its human rights policies and one can only wonder at the level of debate behind the scenes that will be undertaken when it will be Burma's Generals' turn to Chair ASEAN in 2006 (Aung-Thwin, 2003). In the public arena, however, questions have been raised about this issue.

In the US on July 15 2003, the House of Representatives voted unanimously to pass the *Burma Freedom and Democracy Act 2003* and impose a trade blockade on Burma in protest at the crackdown on the NLD and the detention of Aung San Suu Kyi. These sanctions were renewed in May 2004 when the Bush Administration, under international pressure over revelations of prisoner abuse in Iraq, prolonged the restrictions for a further twelve months (Burma junta criticised over talks, 2004, May 18). In 1992, the US bought US$356 million worth of goods from Burma and the loss of this income would have a significant impact on the economy (Borger, 2003, July 16). The US Congress has also consistently requested the Burmese military regime to release Aung San Suu Kyi and other political prisoners (Sein Win, 2004,

January 29). The EU has also threatened to impose sanctions, leaving Australia isolated as one of the very few Western countries maintaining normal relations with the Burmese regime.

In terms of international investment, major energy projects plainly support the Burmese regime and the more companies that leave the country, the more likely the junta is likely to feel pressure to improve its human rights record. It is, however, only pressure from NGOs on governments and corporations around the world that will encourage concrete action. Despite a denial from Premier Oil that pressure from human rights groups had influenced its decision to quit the Yetagun pipeline project in September 2002, it no doubt played a role (Macalister, 2002). Amerada Hess, which until that time had a 25 per cent stake in Premier Oil, was certainly happy with the decision. Vice President, Carl Tursi, admitted that they were happy to have the human rights issue in Burma "removed from the [board room] table" and that it "was certainly a consideration" in investment decisions (Parker, 2002).

Conclusion

The Burmese military junta has an appalling record of human rights relating to forced labour, forced relocation, torture, extrajudicial killings and systematic rape and sexual assault. These human rights violations are often partnered with associated environmental degradation such as destruction of villages, crops and forests, and both can be described as earth rights violations. These violations have often been undertaken in relation to large-scale energy projects such as construction of the Yadana gas pipeline from the Andaman Sea off the Burmese coast to Ratchaburi in Thailand. As partners in this pipeline, the multinationals, Unocal and Total, are complicit in these violations, as is the Thai government which is a majority shareholder in the Petroleum Authority of Thailand (PTT).

The argument that this pipeline is needed for Thailand's 'energy security' is at best fallacious, with the electricity produced from the pipeline being totally unnecessary, as Thailand has an energy glut that will remain for many years due to a continual overestimation of its energy needs. The revenues from the gas purchased by Thailand, however, give crucial support to the

Burmese military junta, both economically and politically, and therefore legitimise the repressive actions of the *Tatmadaw*, the Burmese military, with regard to the ethnic minorities in the pipeline area.

The Thai government itself, led by Thaksin Shinawatra, has become somewhat of a human rights pariah following the extrajudicial killings of at least 2,500 alleged drug dealers early in 2003 and its ongoing aggressive dealings with any NGOs within Thailand who protest against energy projects. Far from providing a contrast to the Burmese situation, the government seems intent on following in its wake.

The *Alien Tort Claims Act* 1789 may provide some remedy for the long-suffering minorities repressed during construction of the Yadana pipeline and may yet set a ground-breaking precedent requiring US companies to consider more carefully the regimes they do business with around the world. In the meantime, only further political pressure from foreign governments, including harsher sanctions such as those imposed by the US government in July 2003, as well as continuing agitation from international human rights groups and other NGOs, will alleviate the repression of those whose misfortune it is simply to live in the path of large-scale energy projects in Thailand and Burma.

References

Ad hoc Commission on Depayin Massacre (2003), 'Preliminary Report of the Ad hoc Commission on Depayin Massacre (Burma)', *Legal Issues on Burma*, no.15, August, pp.1-62.

The Alien Problem, (2003, June 19), *The Economist*.

Apple, B. & Martin, V. (2003), *No Safe Place: Burma's Army and the Rape of Ethnic Women*, Refugees International, April, http://www.refugeesinternational.org/cgi-bin/ri/RapeReport.html

Atrocity in Burma (2003, June 4), *Boston Globe*.

Aung Htoo (2003), 'Depayin Massacre, Crime Against Humanity, National Reconciliation and Democratic Transition', *Legal Issues on Burma*, no.15, August, pp.63-69.

Aung-Thwin, M. (2003), *Hearing on 'Recent Developments in Southeast Asia'*, Subcommittee on East Asia and the Pacific, Committee on International Relations, U.S. House of Representatives, June 10, Washington D.C.

Baker, M. (2003, June 7-8) 'Freedom road delayed and a people betrayed', *Sydney Morning Herald*, p.22.

— (2003, June 21-22), 'ASEAN happy to swallow Burmese regime's lies', *Sydney Morning Herald*, p.18.

— (2004, June 5) 'Burma's road to democracy is no laughing matter.' *Sydney Morning Herald.* http://www.smh.com.au/articles/2004/06/04/1086203633176.html?oneclick=true

Balan Moses (2003, July 9), 'Thaksin: Create strategic alliances', *New Straits Times*.

Bhumiprabhas, S. (2003, May 28), 'Special 1: Law, Order and a Climate of Fear', *The Nation*, http://www.nationmultimedia.com/specials/humanrights/index_special1.php

Borger, J. (2003, July 16), Suu Kyi's plight prompts US sanctions against Burma, *Guardian Unlimited*, http://www.guardian.co.uk/burma/story/0,13373,998917,00.html.

Bureau of International Labor Affairs (1998), *Report of Labor Practices in Burma*, September, (Washington).

Burma junta criticized over talks (2004, May 18), *BBC News*, http://news.bbc.co.uk/2/hi/asia-pacific/3723841.stm

Bush, G. (2003), *Burmese Freedom and Democracy Act: Statement by the President*, Press Release. http://www.burmaproject.org/072803freedomdemocracyact2003.html

Chimprabha, M. (2003, May 28), 'UN envoy cites climate of fear', *The Nation*, http://www.nationmultimedia.com/specials/humanrights/index_may28.php

Christmann, T. (2000), 'The "Unocal Case": Potential Liability of Multinational Companies for Investment Activities in Foreign Countries', *Southern Cross University Law Review*, 4, pp.206-224.

Deighton, M. & Garkawe, S. (2000), 'Confronting the Ghosts of [Burma]'s Past: Rites of Passage for a Democracy', *Southern Cross University Law Review*, 4, pp 169-205

Devraj, R. (2004, June 8). 'Gas pipelines from neighbours to India not pipe dreams.' *Mizzima.com.* http://www.mizzima.com/archives/news-in-2004/news-in-jun/08-jun04-10.htm

Doyle, T. (2004), An Agenda for Envronmental Security in the Indian Ocean Region, in D. Rumley & S. Chaturvedi (eds.), *Geopolitical Orientations, Regionalism and Security in the Indian Ocean*, (South Asian Publishers, New Delhi).

EarthRights International & Southeast Asian Information Network – ERI-SEAIN, (1996), *Total Denial*, (EarthRights International & Southeast Asian Information Network, Bangkok).

Egat pushes for 13% increase in base tariff (1998, Sept 19), *Bangkok Post*, Business, p.1

Egat signs B9.3 billion loan deal to finance major projects (1998, Sept 30), *Bangkok Post*, Business, p.10.

Ekachai, S. (2000, June 7), 'Apathy exacts a high cost', *Bangkok Post*, http://search.bangkokpost.co.th/bkkpost/2000/bp2000_jun/bp200 00608/080600_news19.html

—— (2003, January 16), 'Refusing to give peace a chance', http://search.bangkokpost.co.th/bkkpost/2003/jan2003/bp200301 16/news/16jan2003_news46.html

—— (2003, June 26), 'There's no gain without pain', *Bangkok Post*, http://www.bangkokpost.com/News/26Jun2003_news35.html

Environmentalists turn to religion to save doomed forest (1997, Nov 9), *Bangkok Post*.

Fahn, J (1998, Jan 21), 'Activists toy with builders', *The Nation*.

Garcia, D. (1998), 'Unocal defends its work in Burma – Jed Greer responds', in 'Letter Forum', *The Ecologist*, Vol.28. No.4.

Giannini, T. (1999), *Destructive Engagement: A Decade of Foreign Investment in Burma*, (EarthRights International, Bangkok).

Giannini, T., Redford, K., Apple, B., Greer, J. & Simons, M. (2000), *Total Denial Continues*, (EarthRights International, Bangkok).

Girion, L. (2003, June 15), Pipeline to Justice: Part 1, *Los Angeles Times*.

—— (2003, June 16), Pipeline to Justice: Part 2, *Los Angeles Times*.

Greer, J. (1998), 'US Petroleum Giant to Stand Trial for Burma Atrocities', *The Ecologist*, Vol.28, No.1.

Greer, J. & Giannini, T. (1999), *Earth Rights: Linking the Quests for Human Rights and the Environment*, (EarthRights International, Washington).

Htoo, N., Maung, S., Kher, O., Myo, M.N., MacLean, K., Imamaura, M. & Giannini, T. (2002), *We are not Free to Work for Ourselves: Forced Labor and Other Human Rights Abuses in Burma*, (EarthRights International, Chiang Mai).

IMF to meet amid doubts on future (1998, Sept 28), *The Nation*, p.B5.

International Labour Organization (1998), *Report of the Commission of Inquiry appointed under Article 26 of the Constitution of the International Labour Organization to examine the observance by Myanmar of the Forced Labour Convention 1930 (No.29)*, July 2, (Geneva).

Janchitfah, S. (2002, May 27), 'Pipeline protesters are not done yet', *Bangkok Post*.

—— (2003, January 5), 'Enemies of the State?', *Bangkok Post*.

Kaen, K. (2003, 3 June), 'Thaksin urges quick return to normalcy', *Bangkok Post*,http://www.bangkokpost.com/030603_News/03Jun2003_ news08.html

Karen Women's Organisation (2004), *Shattering silences: Karen women speak out about the Burmese Military Regime's use of rape as a strategy of war*. http://www.ibiblio.org/obl/docs/Shattering_Silences.htm. April.

Killing what is its job to protect (1998, Sept 17), *Bangkok Post*, p.11.

KNU, junta agree on ceasefire after week of talks in Rangoon (2004, January 23), *Bangkok Post*.

Kongrut, A. (2003, June 26), 'Scholars urge PM into talks', *Bangkok Post*, http://www.bangkokpost.net/News/26Jun2003_news05.html

—— (2003, June 30), 'Children call on police to return home', *Bangkok Post*.

Laird, J. (2000), Money Politics, Globalisation, and Crisis: The Case of Thailand, (Graham Brash, Singapore)

Laroche, B. (1996), *Report on Thai/Burmese Gas Pipeline* (FIDH, New York)

Layton, R. (2000), 'Forced Labour in Burma: A Summary of the International Labour Organization Report & Subsequent Developments', *Southern Cross University Law Review*, 4, pp.148-168.

Macalister, T. (2002, September 17), 'UK Group caves in to rights campaigners but claims quitting was expedient', *The Guardian*.

—— (2003, March 26), 'Burma's military halts Premier exit', *The Guardian*.

Mariner, J. (2003), 'Ashcroft's Justice, Burma's Crimes and Bork's Revenge', *FindLaw*, http://writ.news.findlaw.com/mariner/20030526.html

Markels, A. (2003, June 15.), 'Showdown for a Tool in Human Rights Lawsuits', *New York Times*.

Mathiason, N. (2004, January 18), 'Banks bust Burma trade ban', The Guardian, http://www.guardian.co.uk/burma/story/0,13373,1125476,00.html

Myanmar leader returns from ASEAN Summit Meeting (1998, December 17), *Xinhua*.

Myo Lwin (2004, January 19-25), 'Massive natural gas reserve confirmed in Rakhine State', *Myanmar Times*, http://www.myanmar.com/myanmartimes/MyanmarTimes10-200/003.htm

Myo, M.N., Imamura, M., Foley, J., Robinson, N., Maung, S., Htoo, N, & Giannini, T., (2003a), *Entrenched: An Investigative report on the Systematic Use of Forced Labor by the Burmese Army in a Rural Area*, (EarthRights International, Chiang Mai).

—— (2003b), *Supplemental Report*, (EarthRights International, Chiang Mai).

Naw Seng (2002), 'Australia to Assist Judges', *The Irrawaddy*, June, http://www.irrawaddy.org/news/2002/june06.html

Parker, S. (2002, September 17.), 'Activists laud Hess, Premier exit from Myanmar', *Wall Street Journal*.

Payment for Burmese gas could be deferred by PTT (2000, June 7), *Bangkok Post* http://search.bangkokpost.co.th/bkkpost/2000/bp2000_ un/bp20000607/070600_business20.html

Pipe dreams become nightmare for Thailand (1998, Sept 13), *South China Morning Post*.

Power Play (1998, Sept 20), *Bangkok Post*, Perspective, p.2.

Praiwan, Y. (2002, November 16), 'Egat to take part in Salween dam project', *Bangkok Post*.

Pressure builds over Yadana pipeline leak (1998, Sept 30), *Bangkok Post*, http://search.bangkokpost.co.th/bkkpost/1998/september1998/bp 19980930/300998_news14.html

Prospects poor for Burmese gas (1998, Sept 16), *Bangkok Post* http://search.bangkokpost.co.th/bkkpost/1998/september1998/bp 19980916/160998_business04.html

Protestors urge condemnation (2003, June 5), *Bangkok Post*, http://www.bangkokpost.com/050603_News/05Jun2003_news06.h tml

PTT cuts transport cost of Burmese gas to encourage Egat use (2000, Jan 8), *Bangkok Post*, http://search.bangkokpost.co.th/bkkpost/2000/ p2000_jan/bp20000108/080100_business04.html

PTT faces lawsuit over Yadana project (2000, June 4), *Bangkok Post*, http://search.bangkokpost.co.th/bkkpost/2000/bp2000_jun/bp200 00604/040600_news15.html

Puri, Pradeep (2004, January29), 'Myanmar in talks for gas pipeline', *Business Standard*, http://www.business-standard.com/today/ tory.asp?Menu=19&story=33098

Razali puts general in a fix (2003, June 6), *Bangkok Post*, http://www.bangkokpost.com/060603_News/06Jun2003_news05.h tml

Rome Statute of the International Criminal Court (1998), http://www.icc-cpi.int/library/basicdocuments/rome_statute(e).html

Tyranny and terrorism, (2003, June 16), *San Francisco Chronicle*, http://www.sfgate.com/cgi-bin/article.cgi?file=/chronicle/archive 2003/06/16/ED224130.DTL

Kultida Samabuddhi & Yuthana Praiwan (2003, December 18), 'China plans 13 dams on Salween', *Bangkok Post*, p.1.

Sein Win (2004, January 29), 'US asks Burma to release Aung San Suu Kyi', http://www.mizzima.com/archives/news-in-2004/news-in-jan/29-jan04-16.htm

— (2004, February 4), 'DPNS calls on China to reconsider dam project on Nujing', http://www.mizzima.com/archives/news-in-2004/news-in-feb/04-feb04-1.htm

Senators: Open the dam gates (2002, December 20), *Bangkok Post*, http://search.bangkokpost.co.th/bkkpost/2002/dec2002/bp200212 20/news/20dec2002_news07.html

Shan Human Rights Foundation (SHRF) & Shan Women's Action Network (SWAN), (2002), *Licence to Rape*, http:// www.shanland.org/HR/Publication/LtoR/license_to_rape.htm

Slaughter, A.M. & Bosco, D.L. (n.d.), Alternative Justice, http://www.globalpolicy.org/intljustice/atca/2001/altjust.htm.

Tam, P. (2003, June 6), 'Court to Review Whether Unocal Can Be Sued for Alleged Brutality', *Wall Street Journal*.

Treerutkuarkul, A. (2003, 4 June), 'Govt urged to review ties with Burma', *Bangkok Post*, http://www.bangkokpost.com/040603_News/04Jun2003_news16.html

United Nations Commission on Human Rights (2003), 'Situation of Human Rights in Burma, 59th Session, Agenda item 9, April, http://www.ibiblio.org/obl/docs/unchr2003-res.html

United Nations General Assembly (1948), *Universal Declaration of Human Rights*, resolution 217 A(III) of 10 December.

Woods, A. (2003, June 19), 'Suu Kyi held at Myanmar's Insein Jail', *FindLaw*, http://news.findlaw.com/ap_stories/i/1104/6-19-2003/20030619134503_59.html

World Commission of Dams (2000), *The Report of the World Commission of Dams*, http://www.dams.org/report/

Yoon, S. (2003, June 26), 'Thai Talk: Shut up – you're undermining state security', *The Nation*, http://203.150.224.53/page.news.php3?clid=11&id=16135&usrsess=1

Personal Communications

U Aung Htoo (Burma Lawyer's Council General Secretary) (2004), *Interview*, January 19, Mae Sot, Thailand.

Chotirosseranee, Phinan (Co-President, Kanchanaburi Environment Group) (1998), *Interview* (with Co-President and other Members of the KEG), Ellen Cowhey interpreting, October 5, Children Village School, Kanchanaburi Province, Thailand.

Giannini, Tyler (Co-Founder and Co-Director, EarthRights International) (2000), *Interview*, January, Bangkok, Thailand.

—— (2003), *Telephone interview*, June 30.

—— (2004), *Interview*, January 21, Chiang Mai, Thailand.

Ka Hsaw Wa (Co-Founder and Co-Director, EarthRights International) (2004), *Interview*, January 14, Chiang Mai, Thailand.

U Myint Thein (NLD-LA General Secretary) (2004), *Interview*, January 19, Mae Sot, Thailand.

Pipob Udomittipong (EarthRights International/KEG) (2003), *Telephone interview*, June 30.

—— (2004), *Discussions*, January 14, Chiang Mai, Thailand.

Redford, Katharine (Co-Founder and Co-Director, EarthRights International) (2004), *Interview*, January 15, Chiang Mai, Thailand.

Sivaraksa, Sulak (Co-Founder, International Network of Engaged Buddhists) (1998a), *Interview*, January 25, Schumacher College, Devon, UK.

—— (1998b), *Discussions*, September, Bangkok, Thailand.

NOTES

1. The Burmese military junta changed the name of the country from Union of Burma to Union of Myanmar in 1989. Some organisations, such as the United Nations and Amnesty International – which follows UN protocols - have adopted the new usage while others, including Human Rights Watch and exiled organizations such as the Burma Lawyer's Council (BLC), still use the traditional name. Although aware that the name 'Myanmar' has less colonial connotations, the author prefers to use the term 'Burma' until a democratically elected government ratifies the name change.

2. The author would like to gratefully acknowledge the cooperation of *EarthRights International* and the financial support of the Adelaide University Walter and Dorothy Duncan Trust in the research and presentation of this paper.

3. Atrocity in Burma (2003, June 4), *Boston Globe*.

4. The class of people who gain in this situation, usually politicians and their associates, aspire to power largely for the extraction of wealth from the public purse and have come to be refered to collectively as a *kleptocracy*, based on the term kleptomania: an irresistible tendency to steal (Laird, 2000, 10).

5. Although not always called earth rights by these organisations, they received, along with EarthRights International and four other groups, grants from the Richard and Rhonda Goldman Fund for increasing the cooperation between human rights and environmental organizations in recognition of the linkages between the issues (Greer and Giannini, 1999, 94).

6. While this chapter does not allow for a detailed description of the concept of earth rights, the book, *Earth Rights*, by Greer and Giannini (1999) gives a thorough introduction to the legal grounding and different constituent parts that comprise the concept of earth rights.

7. In 2004 Ka Hsaw Wa won the Sting and Trudie Styler Award for Human Rights and Environment, awarded by the Whitley Laing Foundation (http://www.whitley-award.org/News/WhitleyAwards 2004.html). In 1999, Ka Hsaw Wa was a recipient of the Goldman Environmental Prize (http://www.goldmanprize.org/), the Reebok Human Rights Award (http://www.reebok.com/x/us/humanRights /awards/) and the Conde Nast Environmental Prize for his activities highlighting the plight of the ethnic minorities within Burma.

8. 8 It should be noted that the Australian Government has refused to place a ban on commercial links with Burma. The Foreign Minister, Alexander Downer, argues that due to Australia's limited trade with Burma such an action would be little more than symbolic (Baker, 2003, June 21-22). It should be noted that it was such 'symbolic' acts

that successfully brought down the Apartheid regime in South Africa. In addition, the Australian Government's aid policy toward Burma of judicial training and human rights workshops has been roundly criticised by exiled politicians and the Burma Lawyer's Council as having a lack of transparency and conferring legitimacy on the military regime and it's appointed judiciary (Naw Seng, 2002; U Myint Thein, 2004, pers. comm.).

9. These investors include MMA Praxis, the University of Minnesota, a Methodist Church (UK) and a Danish Pension Fund. In addition, Compaq and the Swedish telecom giant, Ericsson, have both withdrawn their operations from Burma over concerns over human rights.

10. It is interesting to note that this was the same tactic the U.S. Army employed in its 'strategic hamlets' relocation program in Vietnam.

11. 11. Hereafter 'the Unocal case'.

12. There is also a similar tort case taking place under California state law with Burmese plaintiffs alleging Unocal was liable for their injury in relation to construction of the pipeline. On January 23, 2004, Superior Court Judge Chaney ruled that Unocal knew that its operations in Burma were likely to result in violations of the human rights of Burmese villagers who were enslaved, killed and tortured in connection with Unocal's natural gas project but that the Unocal subsidiaries in Burma were not the 'alter ego' of the company and Unocal was therfore not liable under the 'alter ego' theory. However Judge Chaney did not strike down the possibility of continuing the trial under two other theories of liability. The trial continues (http://www.earthrights.org/news/ unocaljan04.shtml).

13. After a front page article in the Bangkok Post discussing China's plans for 13 dams on the Salween River upstream from Burma, ERI reported that this publicity had helped activists in China who oppose the dams. This publicity would not have been possible in the restricted media of Burma (Pipob Udomittipong, 2004, pers. comm.; Kultida Samabuddhi & Yuthana Praiwan, 2003, December 18). Publicity for the call by the exiled Burmese group, Democratic Party for a New Society (DPNS), to suspend the project has also only gained coverage outside the country (Sein Win , 2004, February 4).

14. SPDC protestations that they were not involved with the attack have limited credibility as adequately demonstrated by Aung Htoo (2003). Even if one follows the SPDC's own storyline of the incident the most obvious question to be answered, is how, if they were not involved in the incident, they happened to end up with Aung San Suu Kyi and U Tin Oo - for 'protective coustody' - when they were being attacked by 5,000 rabid USDA members.

15. Despite the agreement of a ceasefire between the SPDC and the largest ethnic minority group in Burma, the Karen National Union

(KNU), in January 2004 there were continuing skirmishes between the KNU and the *Tatmadaw* at the time of the meetings in Rangoon, and the agreement is seen, in any case, as an ongoing game by the SPDC of alternate repression of the ethnic minorities and the NLD in order to split the opposition (U Myint Thein, 2004, pers.comm.; KNU, junta agree on ceasefire after week of talks in Rangoon, 2004, January 23). It is currently the NLD who are repressed and peace overtures are being made to the KNU.

CHAPTER 14

Towards an Indian Ocean Energy Community? Challenge Ahead

Sanjay Chaturvedi and Dennis Rumley

Looking ahead, the question is not whether the Indian Ocean region would remain central to the emerging global geopolitics of energy resource competition. It most definitely will. However, as suggested by various contributions to this volume, what remains rather unclear is the extent to which the problematic of who gets what, when, *where* and how will be addressed through diplomacy and dialogue by the various stake-holders in Indian Ocean energy security. Such a probing acquires additional importance in view of the fact that attempts to 'securitise' energy supplies from the Indian Ocean region continue to be made through the threat and use of military power. Indeed, "the strategic military posturing of the US in the Arabian Peninsula, the maritime deployment of US-led Multinational Force enforcing UN sanctions on Iraq, as well as the military occupation of Iraq and the deployment in Central Asia give to the geopolitics of oil in this region a strong military tone" (Billon, 2004, 4).

Furthermore, the attempt on the part of hegemonic power(s) to hide the ruthless pursuit of their so-called 'vital' interests of resource geopolitics behind the discourse and rhetoric of disciplining places and peoples in the name of Western-style democracy, has unleashed a process that insists on creating winners and losers with conflicting political agendas. Marginalised as well as trivialised by such 'resource wars' (ibid.)

are the perspectives and voices that nevertheless dare to challenge the 'new' strategic geography of corporate-style war making as well as the maps that are drawn more on the basis of resource concentrations than political boundaries.

Our key intention in this Chapter is to critically reflect on the broader and deeper context within which the multi-faceted challenge of realising an Indian Ocean Energy Community could possibly be placed and approached on the evolving Indian Ocean research agenda. This context, we argue, is being constituted by the fundamental incompatibility and growing conflict between two paradigms of 'securing' growing energy needs of humankind. The first, and the dominant, paradigm is marked by neoliberal geopolitics which provides a pseudo-scientific rationale to unilateral assertions by the one and only superpower to securitise energy supplies. The second, relatively marginalised, paradigm aims at ensuring energy security from below; reflecting a mindset that upholds the principles of comprehensive security and shows due regard for human rights, gender justice and ecological sustainability.

Securitisation of Energy Flows: Maps of Neo-liberal Geopolitics

It needs to be acknowledged at the very outset that energy security, unlike other aspects of non-traditional security, has always been intimately related to military security (Dietl, 2004). More often than not, it is the hegemonic consumer-states that have sought to maintain an uninterrupted supply of energy at an affordable price, through the threat and/or actual use of military power. It is equally useful to be reminded of the fact that, "after 'black gold' was discovered in Persia in 1908, this resource drastically exacerbated the stakes in the struggle over the spoils of the Ottoman Empire and the Western security imperatives to prevent the (re)emergence of a powerful regional rival" (Billon and Khatib, 2004, 109).

With the energy-military security nexus peaking and asserting itself nearly globally, the prospects of a serious conflict in the Indian Ocean region between the imperatives of energy-social security broadly defined, and the quick-fix strategy of 'securitising' energy flows practised by the hegemonic power(s) cannot be dismissed that easily. Those who choose to address the individual and societal dimensions of energy security have been widely criticised on several grounds, especially after the painful events of 9/11 and its aftermath. It is argued, for example, that shifting the focus away from states and military threats to human-social security is desirable only when threats such as the Iraqi weapons of destruction programme cease to exist (Bilgin, 2003). We are further cautioned that, so long as rogue states, weapons of mass destruction and terrorism prevail, such threats to world order will continue to cancel out other loftier ideas.

There is a growing body of evidence to suggest that a significant shift in American geopolitical thinking on the *Heartland-Rimland* equation is already under way (Klare, 2001). In October 1999, the Department of Defence reassigned senior command authority over American forces in Central Asia from the Pacific Command to the Central Command. Throughout the Cold War, for example, Central Asia had been perceived as a peripheral concern, firmly entrenched within the Soviet sphere of influence, and situated on a remote edge of the Pacific Command's main areas of responsibility - China, Japan, and the Korean Peninsula. However, this is no longer the case. The region, which stretches from the Ural Mountains to China's western border, with vast reserves of oil and natural gas lying in and around the Caspian Sea, is being increasingly perceived as an object of the so-called 'New Great Game' (Stulberg, 2005). Since the US forces in the Persian Gulf region fall within the jurisdiction of the Central Command, the assumption of control by the latter over Central Asia implies that a new strategic map of the 'vital interests' of the US and its allies is in the making.

The growing importance of landlocked Central Asia and its resource endowment in mainstream U.S. geopolitical thinking has an important bearing on the larger geopolitical setting of the Indian Ocean region (Mahalingam, 2004). During the Cold War, the areas that mattered the most to military planners were the zones of confrontation between the US and Soviet blocs: central and southeastern Europe and the Far East. After the dismantling of the Soviet empire, however, these areas have lost much strategic appeal for the United States (except, perhaps, for the demilitarized zone between North and South Korea), while other regions – the Persian Gulf, the Caspian Sea basin, and the South

China Sea – are receiving much greater attention from the Pentagon. To quote Michael Klare (2001):

> Behind this shift in strategic geography is a new emphasis on the protection of supplies of vital resources, especially oil and natural gas. Whereas Cold War-era divisions were created and alliances formed along ideological lines, economic competition now drives international relations - and competition over access to these vital economic assets has intensified accordingly. Because an interruption in the supply of natural resources would portend severe economic consequences, the major importing countries now consider the protection of this flow a significant national concern. In addition, with global energy consumption rising by an estimated two percent annually, competition for access to large energy reserves will only grow more intense in the years to come.

Accordingly, 'security officials' have started paying much greater attention to problems arising from intensified competition over access to critical materials – especially those such as oil that often lie in contested or politically unstable areas. As the National Security Council observed in the White House's 1999 annual report on US security policy, "the United States will continue to have a vital interest in ensuring access to foreign oil supplies" (Peters, 2004, 195-196). Therefore, the report concluded, "we must continue to be mindful of the need for regional stability and security in key producing areas to ensure our access to, and the free flow of, these resources" (ibid.). No sub-region of the Indian Ocean looms as large in these geopolitical calculations as the 'Middle East', which, in a recent study, has also been termed the 'Eurasian Energy Heartland' (Singh, 2002, 288).

The deployment of the term 'Middle East' to designate the current arena of resource warfare suggests, on the one hand, the historical legacy of imperial specifications of the region. This is a good example of how places continued to be transformed into strategic spaces as a result of imperial mapping – both *old* and *new*. It also reminds us of the earlier British imperial designations of the world which were more or less retained on both the wall maps and the mental maps of the intellectuals and institutions of statecraft and foreign policy.

More recently, however, a new term, "Greater Middle East" has entered the realm of competing geopolitical imaginations of *new* Heartlands of the 21st century (Harkavy, 2001, 37). The Greater Middle East, occupying a 'pivotal' position at the juncture of Europe, Africa and Asia, is described as the 'sum' of the core Middle East, North Africa, the African Horn, South Asia and ex-Soviet Central Asia. It is said to be occupying:

> . . . a crucial position with respect to some of the major issue areas of the contemporary era. Those issue areas are energy sources and availability; *the proliferation of weapons of mass destruction (WMD) and their delivery systems; and the dangerous pairings* involving Israel and the Arabs, Iran and Iraq, and India and Pakistan. Surely, this region in its aggregate has come to be viewed by the contending and aspiring world powers – the United States, Russia, a united Europe, China – as a strategic prize, may be *the* strategic prize (ibid.). (emphasis given)

What appears to have emerged from the discussion thus far is that the old cartography of the Cold War is being reassessed and revised by the Pentagon. If this is the case, then the question that demands and deserves our serious consideration is this: who are the makers of these 'new' maps of energy (in)securities? How much continuity and/or change do they represent in terms of defining and defending opportunities and 'threats' before the West? How realist or liberal (or both!) is the agenda behind these new maps? Where, how and why does the Indian Ocean figure on these maps?

With American imperialism articulating itself through global market dynamics of neo-liberal capitalism on the one hand, neo-liberal ideologies and norms are helping make imperial dominance thinkable and doable. It has also been argued forcefully that the geo-economic script written by neo-conservative intellectuals and institutions of statecraft comprises in fact a kind of neo-liberal geopolitics in which the argument for intervention is made in the name of connecting disconnected parts of the globalised world system (Dalby, 2003).

One of the better known protagonists of neo-liberal geopolitics is the US Department of Defence assistant, Thomas Barnett; the author of a widely cited *Esquire* magazine article on 'The Pentagon's New Map' and the architect of the so-called war-map games (Barnett, 2003). The 'compelling' geopolitical script that he has created and delivered to several thousand high-level US government officials was expanded and published as *The Pentagon's New Map: War and Peace in the Twenty First Century* (2004) and quickly became a *New York Times* bestseller.

In Barnett's geopolitical imagination, the USA has now assumed the role and responsibility of the global 'systems manager', who, on a daily basis, has to apply the new global security maxim that 'disconnection defines danger' (Kennelly, 2003). According to him, America stands at an historic crossroads characterised not only by dangers but also by the promise that globalisation may be expanded from a closed club of prosperous states to a planet-wide reality. To him, such a future is worth creating and US 'security exports' can actually make it happen by disciplining the 'real' to fit into the imagined.

The Pentagon's 'New' map is being drawn around much of the world's mid-section, encircling the regions from where threats to national and global security are likely to emanate, namely the *Non-Integrating Gap*. The Gap stretches from the Andean states of Latin America and the Caribbean Basin to sub-Saharan Africa, the Middle East and Central Asia, and much of Southeast Asia. In Barnett's view, the Gap is 'dangerous' because it is disconnected from the institutions and infrastructure of economic globalisation. He is of the view, therefore, that the Gap needs to be integrated into globalisation's *Functioning Core* in order to prevent further terrorist attacks like 9/11. This alone, he proclaims, could bring about a durable and just peace. Since disconnectedness is now said to be the 'clear and present danger' and the US and its allies must ensure that the Gap keeps shrinking at all costs. Furthermore, Barnett argues, the US military is now due for a strategic shift. Rather than getting embroiled in wars against great powers, the US should follow an entirely different strategy that anticipates a series of long-term 'aggressive' engagements in the Gap, for the sake of a future worth creating. Involving major changes in every aspect of military organisation, foreign policy, and national strategy, the implications of Barnett's new grand strategy are rather profound and far-reaching.

What Barnett appears to have done is to lay out a new security strategy that the neo-conservative intellectuals and institutions of statecraft in the United States have been waiting for since the end of the Cold War, and even more desperately since 9/11. There are some remarkable similarities between Thomas Barnett and his *The Pentagon's New Map* and George Kennan and his famous X *Article*. Whereas the latter revealed the grand strategy of Soviet containment in the aftermath of World War II, the former attempts to provide a an equally grand strategy of connecting the places and peoples that remain disconnected to the agenda of US dictated corporate globalisation in the aftermath of the Cold War. As succinctly put by Susan Roberts, Anna Secor and Matthew Sparke (2003, 888-89):

> In the neo-liberal approach, the geopolitics of inter-imperial rivalry, the Monroe doctrine, and the ideas about hemispheric control . . .are eclipsed by a new global vision of almost infinite openness and interdependency. In contrast also to the Cold War era, danger is no longer imagined as something that should be contained at a disconnected distance. Now, by way of a complete counterpoint, danger is itself being defined as disconnection from the global system.

One good example of how complex issues related to energy security in the Indian Ocean region are being implicated within what we have described earlier as the broader and deeper context of neo-liberal geopolitics relates to the manner in which the control of the Iraqi oil sector is furnishing the Bush administration with the opportunity of testing a 'freedom oil' policy. As Philippe Le Billon and Fouad El Khatib put it so forcefully (2004, 127-28):

> Narratives around freedom versus evil have been extensively used by the Bush administration in its discursive construction of terrorism and the justification of the 'war on terror'. The same need to contrast the 'Saddam era' from the (US) 'liberation era' entails that oil comes to play a different role for the population than is associated in the 'oil curse' narrative commonly used to portray the previous Iraqi

regime (and most of those in the region). In other words, oil needs to bring 'freedom' rather than 'evil' in Iraq and the broader region. Such discursive construction entails a number of policies, whereby the US would take an active role in creating and sustaining a political and institutional environment, as well as investment and infrastructure, in which oil and oil revenues consolidate a stable democratic regime and lessen regional tension - thereby also resolving some of these conflicts and dilemmas between energy and broader foreign policy goals.

The aggressive engagement in Iraq is further justified in terms of self-proclaimed 'moral exceptionalism' (Agnew, 1983), ruthlessly pursued 'manifest destiny' and zealously guarded experience of the moral geography of the 'frontier'. The 'democratic' disciplining of the 'failing'/'fading' states of Afghanistan and Iraq through the exercise of military power by the one and only superpower suggests at the same time the performance of power relations understood as 'dominance'. According to Matt Sparke (2004:1-2), the phenomenon of dominance could be defined as, "a particular form of hegemony articulated (and thus both experienced and consolidated) more through coercion than consent. According to Spark, this does not mean that dominance eclipses the ideological interpellation of consent altogether. After all, hegemony works through *both* consent and coercion. But what it does mean is that the overall practice and performance of hegemony is rather skewed in favour of brute force, naked violence and coercion.

Be it hegemony or dominance, or both, various attempts at building regional and sub-regional partnerships/alliances, as building blocks for an Indian Ocean Energy Community (IOEC), discussed at length in this volume, are likely to be undermined by 'coercive Western Energy Security Strategies' and the North-South dimensions of resource conflicts (Peters, 2004). To quote a keen analyst of 'resource wars':

. . . in the future we will be confronted with new resource wars in the international system, which will be precipitated by two developments: first, an anticipated oil supply crisis as a first consequence of the decline of global oil reserves, and, second, the uneven distribution of these declining resources along the North-South axis . . . in response to these developments, the coercive character of traditional US strategies for securing energy will intensify, thus bearing the potential to escalate into further armed conflicts.

> . . . there are only two sustainable strategies for conflict prevention: first the reduction of the dependency on fossil fuels by developing alternative and renewable energy, and second, the pursuit of a global policy based on more equitable and controlled energy distribution (ibid., 2004, 187).

The emphasis supplied in the above quotation provides us with an entry point into the discussion of an alternative paradigm of energy security; one that not only calls for a dialogue between the suppliers and the consumers but underlines the importance of placing the notion of energy security within the context of human-ecological security.

Energy Security: Perspectives from Below

For those who believe in the idea of an Indian Ocean community, it is difficult to overlook the ground reality that, by and large, Indian Ocean is a 'Third World' Ocean, sought after by the 'First Worlds' both within and outside the region in the context of resource geopolitics. The region is the home of a vast majority of the poorest of the poor. It is to state the obvious perhaps that for the poor, the priority is the satisfaction of such basic human needs as jobs, food, health services, education, housing, clean water and sanitation. Energy plays a central role in ensuring delivery of most of these services. The Indian Ocean Energy Community has its work cut out for it with hundreds of millions lacking access to electricity, and equally large number of people still relying on traditional biomass fuels.

The local practices throughout the Indian Ocean region, as opposed to the 'global truths' constructed by neo-liberal geopolitical reasoning underlying the 'new' Pentagon map, suggest beyond doubt that deprivation in energy has enormous impacts on the lives of poor people; in other words human security. As aptly argued by Kanti Bajpai (2002, 214-25), "the point of human security studies . . . is to describe a map of violence that goes well beyond the map created by the neo-realist/statist view of security. Evidence suggests that the map is much larger than the map of inter-state violence. With all its imperfections, a human security audit, done systematically and rigorously, will map a massive area of human experience that is presently unmapped (or mapped in bits and pieces). Its promise is not to get every contour absolutely right; it is rather, to start to fill, however incompletely, what is presently a very blurry picture – a picture of a turbulent and complex world system, composed of a ramifying set of actors and linkages, as well as an emergent world society increasingly latticed by globalised norms and institutions".

The energy sector plays a very important role in the social and economic development of a society. It is self-defeating in both the short and long run to deny the vital links that exist between the energy sector and poverty reduction - through income, health, education, gender, and the environment. The energy securitisation paradigm is ideologically blind to the complexity involved in energy security. There is also a wilful denial of the fact that the energy sector has to work closely with other sectors in order to tackle energy deprivation. Easier said than done, eradicating energy deprivation calls for tough public policy choices that include, but do not exhaust, the delicate balancing of diverse interests and sustained commitment. It needs to be emphasised that energy communities can not be imposed from above but are created through co-operative practices from below. After all, how best to design and implement policies for expanding equitable access to energy is a question requiring close attention to the complex details of social hierarchies and power asymmetries within a nation-state.

According to R. K. Pachauri (2005), a distinguished energy security expert and the director of Tata Energy Research Institute (TERI), New Delhi, "energy security and, therefore, the prospects of stable and healthy economic growth, would depend on timely initiatives involving not only energy supply sectors but major energy-consuming sectors too." The following observations by him, in our view, have relevance for the entire Indian Ocean region and not only for India. It could be highly misleading to assume, argues Pachauri, "that securing the oil supply for a particular country's growing demand would be adequate in itself to ensure energy security. An enlightened energy policy demands that, in addition to an assurance of steady economic growth and the ability to withstand future oil price shocks, the very structure of a country's economy should also match the assurance of sustainable and manageable supply options" (ibid.).

In the light of the issues raised above, a question that needs to be addressed with a sense of urgency is this: to what an extent are the actual practices in the IOR dictated and driven by the principles of energy security from below? And, to what extent are such principles institutionalised? A critical assessment of the 'usual' functioning of the transport sector in most of the countries of the Indian Ocean region revels that, with a shift in the share of both passenger and freight transport away from rail to the oil-intensive road transport, it is becoming increasingly oil-intensive.

A threat to a country like India's energy security emanates from the sheer inability of the stagnating Indian railway system to provide service capable of weaning away traffic from road transport. In the absence of inadequate attention being paid to the promotion of adequate and reliable public transport systems, we are more likely to witness an escalating proliferation of personal vehicular transport. Much less satisfactory is the record of inadequate development of inland water transport and coastal shipping in most of the Indian Ocean region. Despite the fact that for every passenger or unit of goods moved, waterborne transport is the most energy-efficient option, which a vast majority of countries in the region have been singularly unsuccessful in utilizing. In short, the Indian Ocean region not only needs an efficient energy supply industry, as well as diverse sources of energy ranging from oil to solar and wind, but more importantly an economic structure which uses energy efficiently. The transport sector, the commercial complexes and residential buildings under construction could easily lock the region into an energy-intensive pattern which goes against the objectives of energy security.

Do we see any signs of an Indian Ocean energy security community as yet, or is it more or less wishful thinking? Even if the ground realities are still waiting to be touched and transformed in a meaningful manner by the widened understanding of energy security, the paradigm of energy-human security seems to have captured the imaginations of several policy-makers in the Indian Ocean region, and hopefully, even beyond.

In his most recent book, Zbigniew Brezezinski (2004) has made a forceful argument in favour of the USA offering 'global leadership' rather than asserting 'global domination' on what he would still perceive as the Eurasian chessboard. According to him, "together, the United States and the European Union represent the core of global political stability and economic wealth" (ibid, 89). The proverbial billion-dollar question facing the United States today, according to Brezezinski, is this: How to strike a balance between sovereign hegemony and an emerging global community, and how to resolve the dangerous contradiction between the values of democracy and the imperatives of global power" (ibid, 138). This, according to him, "remains America's dilemma in the age of globalisation".

The above mentioned work by Brezezinski appears to suggest that, while the dominant security paradigm within the United States is one of 'keeping the violent peace', there is the dawning of a recognition that this might turn out to be increasingly difficult, not least as trends in asymmetric warfare make it exceptionally difficult to maintain control by conventional means. It has also been argued by some analysts that, "the United States cannot and should not resist the end of the American era. To do so would only risk alienating and provoking conflict with a rising Europe and an ascendant Asia . . If armed with the right politics, the United States may well be able to manage peacefully the transition from unipolarity to multipolarity, thereby ensuring that the stability and prosperity attained under its watch will extend well beyond its primacy" (Kupchan, 2002, 247).

While such alternative thinking might at present be limited to just a few far sighted individuals in the military, academic and policy communities, there is potential for this to develop into a substantial intellectual movement that assists in effecting a paradigm shift. Such a paradigm shift is needed both among and within nation-states globally, and not simply among the Indian Ocean states, despite the obvious appeal of both neo-liberal geopolitics and realism. Realism is "easy to grasp because it portrays the world in clear friend-or-foe terms" (Kupchan, 2002, 234). Whereas the appeal of neo-liberal geopolitics lies in its reductionism. Both are hopelessly incapable, however, of ensuring energy security to peoples at large.

Indian Ocean's energy issues, as this volume has shown, are complex as well as diverse in nature. Even though the state and non-state stakeholders in Indian Ocean energy security issues may have competing interests, some analysts are of the view that, "with the market playing an increasingly important role in energy security, economic geopolitics suggests that the costs of rivalry over strategic energy resources and supply are vastly outweighed by the benefits of co-operation . . . in fact, co-operation in energy security issues could pave the way for overall regional security co-operation" (Dadwal, 2002, 83).

Yet there is no guarantee that good economics will automatically be translated into good political practices. Some of the foundation stones of an Indian Ocean Energy Community are being laid down at present in the form of gas pipeline projects being proposed/negotiated among Turkmenistan, Afghanistan, Iran, Pakistan, India, Bangladesh and Myanmar, and China. The success of these projects, some of which have been critically examined in this volume, calls for multilateral diplomatic efforts and multi-sectoral partnerships. The need for a change of mindset as well as institutional reforms in this regard is graphically emphasised in the following excerpt from the address by Mr. Mani Shankar Aiyar, Minister for Petroleum and Natural Gas, Government of India (who has shown an exceptional vision and enthusiasm while proposing an Asian Energy Security Grid) delivered at the Observer Research Foundation, New Delhi, on 16 March 2005 (ORF Energy, 2005):

> It is blindingly clear that it is in everybody's interest to promote energy cooperation in our extended neighbourhood but what is crystal clear to us is not very clear to many foreign offices. This includes our own and it certainly includes those of our immediate neighbours but when we get into the extended neighbourhood their economies take precedence over our economics. So, although there have been some major breakthroughs in recent weeks, particularly I would say the decision of the cabinet on the 9th of February to mandate us to conduct negotiations with our extended neighbourhood and our immediate neighbourhood, the bulk of the problems on the ground are ahead of us and not behind us. So, while I share everybody's delight that we have

been able to make the conceptual breakthrough, I have to emphasize that what we have done so far is no more than taking the initial steps. *Unless everybody is as enlightened in his politics as he is demonstrating in his economics, we are not going to reach our goals as quickly as we need to.* (emphasis supplied).

At the heart of the enlightened politics hinted at above, which no doubt is needed for ecologically sustainable and socially just pursuits of energy security in the Indian Ocean Region, should also be positioned, in our view, the resistance to the discourse of neo-liberal geopolitics and war-making practices that flow from it. The claim of neo-liberal geopolitics that peace and prosperity can only truly prevail in areas where the US has established military bases, and ongoing security alliance also needs to be seriously questioned. Failure to do so is likely to reinforce the prevailing temptations and tendencies to securitize access to and control over the energy resources of the Indian Ocean region.

References

Agnew, J. (1983), "An Excess of National Exceptionalism: Towards a New Political Geography of American Foreign Policy", *Political Geography Quarterly* 2(22), pp.152-66.

Bajpai, K. (2002), "Beyond Comprehensive Security: Human Security" in *Comprehensive Security: Perspectives from India's Regions.* Seminar Proceedings, New Delhi: Policy Research Group. August 2001.

Barnett, T.P.R. (2003), "The Pentagon's New Map", *Esquire*, March. http://www.nwc.navy.mil/newrulessets/The PentagonsNewMap. htm

Barnett, T. P. R. (2004), *The Pentagon's New Map: War and Peace in the Twenty First Century*, Putnam Adult.

Bilgin, P. (2003), "Individual and Societal Dimensions of Security", *International Studies Review*, 5, pp. 203-222.

Billon, P. L. (2004), "The Geopolitics of Resource Wars", *Geopolitics*, 9(1), pp. 1-28.

Billon P.L. and Khatib, F. E. (2004), "From Free Oil to 'Freedom Oil': Terrorism, War and US Geopolitics in the Persian Gulf", *Geopolitics*, 9(1), pp. 108-109.

Brzezinski, Z. (2004), *The Choice: Global Domination or Global Leadership*, New York: Basic Books.

Dadwal, S. R. (2002), *Rethinking Energy Security in India*, New Delhi: Knowledge World.

Dalby, S. (2003), "Calling 911: Geopolitics, Security and America's New War", *Geopolitics* 8(3), pp. 61-86.

Dietl, G. (2004), "New Threats to Oil and Gas in West Asia: Issue in India's Security", *Strategic Analysis*, 28(3), pp.373-389.

Klare, M. (2001), *Resource Wars: The New Landscape of Global Conflict*, New York: Metropolitan Books.

Kupchan, C. A. (2002), *The End of the American Era: U.S. Foreign Policy and the Geopolitics of the Twenty-First Century*, New York: Vinatage Books.

Mahalingam, S. (2004), "Energy and Security in a Changing World", *Strategic Analysis* 28(2), pp. 249-271.

Pachauri, R.K. (2005), "The Black Gold's Curse: Despite Several Myths, Energy Security has become new Global Concern", *Outlook*, New Delhi, 2 May.

Peters, S. (2004), "Coercive Western Energy Security Strategies: 'Resource Wars' as a New Threat to Global Security", *Geopolitics*, 9(1), pp. 187-212.

Roberts, S., Secor, A. and Sparke, M. (2003), "Neoliberal Geopolitics", *Antipode*, 35(5), pp. 886-897.

Rogers, P. (2000), *Losing Control: Global Security in the Twenty-First Century*, London: Pluto Press.

Singh, K.R. (2002), "Geo-strategy of Commercial Energy", *International Studies* 39(3), pp. 259-288.

Sparke, M. (2004), "Political Geography: Political Geographies of Globalization (1) Dominance", *Progress in Human Geography*, 28(6), pp.1-18.

Stulberg, A. N. (2005), "Moving Beyond the Great Game: The Geoeconomics of Russia's Influence in the Caspian Energy Bonanza", *Geopolitics*, 10(1), pp. 1-25.

Index